Machine Learning for the Physical Sciences

Machine learning is an exciting topic with a myriad of applications. However, most textbooks are targeted towards computer science students. This, however, creates a complication for scientists across the physical sciences that also want to understand the main concepts of machine learning and look ahead to applications and advancements in their fields.

This textbook bridges this gap, providing an introduction to the mathematical foundations for the main algorithms used in machine learning for those from the physical sciences, without a formal background in computer science. It demonstrates how machine learning can be used to solve problems in physics and engineering, targeting senior undergraduate and graduate students in physics and electrical engineering, alongside advanced researchers.

Key Features:

- Includes detailed algorithms.
- Supplemented by codes in Julia: a high-performing language and one that is easy to read for those in the natural sciences.
- All algorithms are presented with a good mathematical background.

Carlo Requião da Cunha is currently an assistant professor at the School of Informatics, Computing, and Cyber Systems at Northern Arizona University. He holds a Ph.D. degree in electrical engineering from Arizona State University. Throughout his career, Dr. da Cunha has held various academic positions and research affiliations in institutions such as McGill University, Chiba University, and the Technical University of Vienna. His research focuses on computational science, where he applies machine learning techniques to the design of innovative electronic devices and systems.

Machine Learning for the Physical Sciences

Fundamentals and Prototyping with Julia

Carlo Requião da Cunha

CRC Press
Taylor & Francis Group
Boca Raton London New York

CRC Press is an imprint of the
Taylor & Francis Group, an **informa** business

Designed cover image: Shutterstock_ 131970302

First edition published 2024
by CRC Press
2385 NW Executive Center Drive, Suite 320, Boca Raton FL 33431

and by CRC Press
4 Park Square, Milton Park, Abingdon, Oxon, OX14 4RN

CRC Press is an imprint of Taylor & Francis Group, LLC

© 2024 Carlo Requião da Cunha

ISBN: 978-1-032-39229-5 (hbk)
ISBN: 978-1-032-39523-4 (pbk)
ISBN: 978-1-003-35010-1 (ebk)

DOI: 10.1201/9781003350101

Typeset in Nimbus Roman
by KnowledgeWorks Global Ltd.

Publisher's note: This book has been prepared from camera-ready copy provided by the authors.

Dedication

---

for Arthur, Sofia, and Ingrid

Contents

SECTION I   *Foundations*

SECTION II   *Unsupervised Learning*

SECTION III Supervised Learning

SECTION IV Neuronal-Inspired Learning

SECTION V Reinforcement Learning

SECTION VI Optimization

Preface

This book is the result of several years of teaching courses such as Data Analysis and Computational Methods in a department of Physics at both the undergraduate and graduate levels. The main challenge posed by teaching these courses, as well as by writing this book was how to translate contemporary machine learning approaches, usually conceived by computer scientists and statisticians, to students of physics. The cultures of these three communities are very distinct. For instance, while statisticians prime for proper mathematical formalism, computer scientists are more concerned with developing algorithms and computational solutions. While computer scientists have a conference publishing culture, statisticians usually prefer publishing in journals. Physicists, on the other hand, tend to prefer a high level of mathematical formalism, but not as high as that held by mathematicians and statisticians. Physicists are concerned with applying machine learning concepts to solve problems, but not the same types of problems faced by computer scientists. On top of that, each community uses a specific notation. Finding a proper balance between those fields and producing a material that would be palatable to physics students was both the major goal and challenge of this book.

The next challenge was regarding the title of this book. The term "machine learning" carries, to many, the notion that a machine is capable of learning in the same way a human does. A statistician would probably argue that this tries to anthropomorphize machines, and many methods in the field have been used by statisticians (and physicists) for decades. A computer scientist, on the other hand, could counter-argue by saying that we do not know exactly how the human brain works and we might be getting close to developing computers that "learn". The consensus is, however, that the term "machine learning" is already popularized and relates to a set of algorithms that adjust to data. Therefore, the title of the book is "machine learning" in the sense that it is about algorithms that adjust to data, but it also clearly states that it is "for the physical sciences". Do not expect to use this book to learn how to develop a new AI language-model-based chat box, but rather how to use some of the most popular machine learning algorithms to solve problems in physics.

This leads to the third challenge found when writing this book. Although this book is about machine learning, it does not aim at teaching students how to code and use libraries popularized in the computer science community. Although this book is about physics, it does not aim at teaching students how to solve particular problems in physics. The main goal was to translate many machine learning algorithms into a language that physicists can understand and use. To achieve this goal, a combination of discussions, mathematical formalism, and coding was used. The language chosen for the codes was Julia for two reasons: i) it is a very readable language that maps mathematics to code very clearly, and ii) the execution of its programs can be significantly faster than that of many other languages, and this is particularly necessary for solving problems in physics.

"Everything was made as simple as possible, but not simpler."[a] Therefore, snippets are short and do not include fancy calculations. They are straight to the point, written with the main objective of illustrating the theory behind them, not producing the most efficient algorithms.

The machine learning research community is highly active and produces a significant number of papers every week. Although it was impossible to incorporate the newest developments in this book, I tried to incorporate the most used algorithms. Because of their complexity and length, some algorithms may be presented in an algorithmic form instead of code, but codes for all algorithms are provided outside the book. Finally, this book is an introduction to an active and fast-developing field. The finite number of pages limits the depth of the discussions, but references for every topic are provided, and the readers are encouraged to explore them.

Finally, like many of my works, this book has been written employing three languages: English, Mathematics, and Julia. Just as proficiency in English reading is essential, so is the ability to comprehend mathematical notation.

Structure

This book is divided into six sections. Section I is about the foundations necessary to understand machine learning. Here, multivariate calculus and probability theory are discussed. Topics such as tensors, matrix calculus, and affine maps are introduced in the multivariate calculus chapter. The chapter about probability theory deals with modern concepts such as measure theory, information theory, kernel estimation, and Markov chains.

Section II is about unsupervised learning, or algorithms that work with unlabeled data and try to find patterns and structures within this data. Topics include dimensionality reduction using techniques such as principal component analysis, cluster analysis, and vector quantization techniques.

Section III deals with supervised learning, or algorithms that "learn" how to map inputs to outputs given some training data, which can often be labeled. The two main classes of supervised algorithms are covered: regression and classification models. In the chapter about regression models, linear, non-linear, and Gaussian regressions are covered. In the chapter about classification, logistic regression, trees, support vector machines, and the naïve Bayes algorithms are discussed.

Section IV covers neuronal-inspired learning, or algorithms that draw inspiration from biological neurons and neural networks in the brain. One chapter in this section discusses feedforward neural networks and autoencoders including optimization strategies to obtain high training efficiencies. Another chapter is dedicated to advanced architectures including the convolutional neural network, widely used in computer vision, recurrent networks, and the Hopfield and Boltzmann networks.

Section V is about reinforcement learning, or algorithms used for agents to make sequential decisions when navigating in an environment. These algorithms have become popular in applications ranging from game playing to autonomous driving. One

---

[a] A famous quote attributed to Albert Einstein.

chapter discusses two very popular model-free algorithms: SARSA and Q-Learning. Another chapter is dedicated to reinforcement learning algorithms that rely on the gradient of a reward function: REINFORCE, the actor-critic network, and trust region methods.

Finalizing this book, section VI deals with optimization strategies that can be used, for example, to train neuronal-inspired algorithms, or to design electrical circuits. First, we discuss population-based metaheuristic methods such as genetic algorithms and particle swarm optimization. Then we discuss local search methods such as the popular amoeba optimization and simulated annealing widely used in statistical mechanics.

Carlo Requião da Cunha
Flagstaff, AZ
July 2023

List of Figures

List of Tables

Section I

Foundations

1 Multivariate Calculus

Before diving into the study of machine learning, it is important to establish a strong foundation with clear definitions that will help us understand and work with machine learning algorithms and methods. For example, it is common for students to confuse matrices, tensors, and multidimensional arrays. In this chapter, we will examine each of these fundamental mathematical objects and explore how to perform operations with them.

Our first step will be to define these terms in detail[a]. Later, we will introduce the concept of spaces and mapping between these spaces, which play an important role in topics such as dimensionality reduction. Toward the end of the chapter, we will discuss matrix calculus, which is a powerful mathematical tool for optimizing machine learning models that use techniques such as gradient descent and backpropagation.

1.1 VECTORS, MATRICES, AND TENSORS

A field F is defined as a set equipped with two binary operations $F \times F \to F$, namely addition and multiplication, such that for any a, b, and c being elements of this field, they satisfy the properties of:

Associativity: $a + (b + c) = (a + b) + c$ and $a \cdot (b \cdot c) = (a \cdot b) \cdot c$

Commutativity: $a + b = b + a$ and $a \cdot b = b \cdot a$

Distributivity: $a \cdot (b + c) = (a \cdot b) + (a \cdot c)$.

Also, there are two special elements:

Identity: $a + 0 = a$ and $a \cdot 1 = a$

Inverse: $a + (-a) = 0$ and $a \cdot (a^{-1}) = 1$.

One example of a field is the real numbers $\mathbb{R}$ equipped with the operations of addition and multiplication.

Vectors are elements of a *vector space*[b]. A vector space over a field F is a non-empty set V equipped with the operation of *vector addition* that takes two vectors to a new one $V \times V \to V$ and the operation of *scalar multiplication* (not to be confused with the *scalar product*) $F \times V \to V$ that takes an element a of a field F and a vector $\mathbf{v} \in V$ to a vector $a\mathbf{v} \in V$. If $F \subseteq \mathbb{R}$, then a scales the vector $\mathbf{v}$ and is consequently called a *scalar*.

A 2-element vector in Julia [5] is given by:

```
v = [1,2]
```

---

[a] Good references for this whole chapter are [3,4].

[b] A space is a set of mathematical objects (numbers, functions, geometric objects, or anything that can be formally defined), called points, and a collection of relationships between these objects.

Table 1.1

Vectors, Matrices, and Tensors

x	scalar
$\mathbf{x}$	vector represented by a column matrix
$\mathbf{x}^{\top}$	vector represented by a row matrix
X	field or set
$\mathbf{X}$	matrix or multidimensional array
X_{ij}	element of a matrix
X_{ijk}	element of a multidimensional array

A matrix is a rectangular array with n rows and m columns such that its elements are mathematical objects. We will use bold uppercase letters to denote matrices (e.g. $\mathbf{A}$) and regular uppercase letters to describe its elements (e.g. A_{ij} indicates the element in the ith row and the jth column). Matrices with a single column or a single row will be used to represent vectors.

A 2-element row matrix in Julia is given by:

```
R = [1 2]
```

whereas a 2-element column matrix in Julia is given by:

```
C = [1; 2;;]
```

Tensors[c] are algebraic objects used to describe multilinear relationships between sets of algebraic objects related to a vector space ($V_1 \times \ldots \times V_N \to W$, where V_n and W are vector spaces). Just as row and line matrices can be used to represent vectors, tensors can be represented by multidimensional arrays.

A $2 \times 2 \times 2$ multidimensional array of 64-bit floating point numbers in Julia is given by:

```
M = Array{Float64}(undef,2,2,2)
```

---

[c]Conceived in 1900 by the Italian mathematicians Tullio Levi-Civita (1873–1941) (adviser) and Gregorio Ricci-Curbastro (1853–1925) (advisee), following the work of Berhard Riemann (1826–1866) German mathematician advised by Carl Friedrich Gauss and Elwin Bruno Christoffel.

To illustrate how multidimensional arrays can represent vectors and tensors, let's consider the following set of linear equations:

$$A_{111}v_1 + A_{121}v_2 = B_{11}$$
$$A_{211}v_1 + A_{221}v_2 = B_{21}$$
$$A_{112}v_1 + A_{122}v_2 = B_{12}$$
$$A_{212}v_1 + A_{222}v_2 = B_{22}$$

$$(1.1)$$

This can be represented by matrices:

$$\begin{bmatrix} A_{111} & A_{121} \\ A_{211} & A_{221} \end{bmatrix} \begin{bmatrix} v_1 \\ v_2 \end{bmatrix} = \begin{bmatrix} B_{11} \\ B_{21} \end{bmatrix}$$

$$\begin{bmatrix} A_{112} & A_{122} \\ A_{211} & A_{222} \end{bmatrix} \begin{bmatrix} v_1 \\ v_2 \end{bmatrix} = \begin{bmatrix} B_{12} \\ B_{22} \end{bmatrix}$$

$$(1.2)$$

This, however, can be simplified if we use multidimensional arrays:

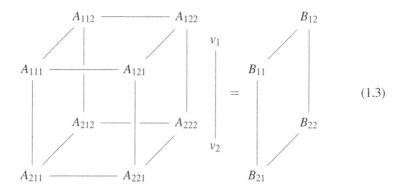

$$(1.3)$$

Note that the second array on the left-hand side of the equation is a vector that can be represented by a column matrix and the array on the right-hand side of the equation can be represented by a matrix:

$$\begin{bmatrix} v_1 \\ v_2 \end{bmatrix} = \begin{bmatrix} B_{11} & B_{12} \\ B_{21} & B_{22} \end{bmatrix}$$

$$(1.4)$$

In index notation this can be written as:

$$A_{ijk}v^j = B_{ik}, \tag{1.5}$$

where we have used Einstein's[d] summation rule[e]. This is a *contraction* operation that is diagrammatically shown in Fig. 1.1. Observe that, in this operation, a rank-3 tensor contracts with a vector producing a rank-2 tensor. In general, contractions between a D-dimensional object and a d-dimensional object produce a $D+d-2$ object since one index from each object is removed. In this particular case, we have $D = 3$ and $d = 1$, producing $3 + 1 - 2 = 2$.

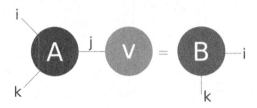

Figure 1.1 Contraction between a rank-3 tensor **A** and a vector **v** producing a rank-2 tensor **B**

> **Basis**
>
> Consider a vector space U over some field F. The basis B of U is the smallest set of linearly independent vectors $\{\hat{\mathbf{e}}_i\}$ that contains U. This means that any vector $\hat{\mathbf{e}}_i \in B$ cannot be written as a linear combination of the remaining *basis vectors*. Moreover, any vector in U can be written as a linear combination of these vectors. The coefficients of this linear combination are known as the *coordinates* of such a vector.

Tensors are not multidimensional arrays, though multidimensional arrays may represent a tensor. Tensors represent transformations. For instance, a vector must be invariant to the coordinate system. Therefore, for two different basis $\{\hat{\mathbf{e}}_i\}$ and $\{\hat{\mathbf{f}}_j\}$, a vector **v** can be written as:

$$\mathbf{v} = v^i \hat{\mathbf{e}}_i = u^j \hat{\mathbf{f}}_j, \tag{1.6}$$

If the magnitude of $\hat{\mathbf{f}}_j$ is twice the magnitude of $\hat{\mathbf{e}}_i$, then the component u^j will have to be half of v^i so that **v** is still the same. When the components vary contrary to the basis, we say they are *contravariant* (or *tangent vector*).

---

[d] Albert Einstein (1879–1955) German physicist, winner of the Nobel Prize in Physics in 1921.

[e] An index not previously defined that appears twice in a single term indicates a summation of that term overall values of that index. Upper indices represent components of contravariant vectors whereas lower indices represent components of covariant vectors.

If we now take the projection of $\mathbf{v}$ onto itself, we get:

$$\|\mathbf{v}\|^2 = \left(v^i\hat{\mathbf{e}}_i\right)\cdot\left(u^j\hat{\mathbf{f}}_j\right) = \left(\hat{\mathbf{e}}_i\cdot\hat{\mathbf{f}}_j\right)u^jv^i = g_{ij}u^jv^i = u_iv^i. \tag{1.7}$$

Since the norm of the vector is invariant, u_j varies opposite to the contravariant components and consequently in the same way as the basis. Hence, we call them *covariant* (or *covector, one-form, linear form, dual vector,* or *cotangent vector*). Also in the formula, we have $g_{ij} = \hat{\mathbf{e}}_i\cdot\hat{\mathbf{f}}_j$ which is called a *metric tensor*. To clarify, contravariant vectors reside in one vector space, while covariant vectors exist in a *dual space*. It is important to note that despite their distinct locations, both types represent the same underlying vector.

Let's now take the derivative of Eq. 1.6 for u^j:

$$\hat{\mathbf{f}}_j = \frac{\partial v^i}{\partial u^j}\hat{\mathbf{e}}_i. \tag{1.8}$$

Since the new basis is defined as a linear combination of the old basis, we say that this is a *covariant transformation*. This transformation can be identified with lower indices in the entities that are being transformed. Conversely, we can take:

$$du^j = \frac{\partial u^j}{\partial v^i}dv^i. \tag{1.9}$$

Since vector components transform opposite to the basis, we say that this is a *contravariant transformation* and this can be identified with the upper indices.

The differential form of Eq. 1.7, known as the *first fundamental form* is given by:

$$(d\mathbf{s})^2 = g_{ij}dv^idv^j. \tag{1.10}$$

Coupling this with Eq. 1.9, we get:

$$(d\mathbf{s})^2 = g_{ij}\frac{\partial v^i}{\partial u^k}\frac{\partial v^j}{\partial u^l}du^kdu^l. \tag{1.11}$$

This gives the square of the arc length $(d\mathbf{s})^2$ in the basis of the transformed coordinates.

Example 1.1.1. Change of coordinates

Let's take a vector in $\mathbb{R}^2$ and represent it in a new coordinate system corresponding to an amplified rotation given by:

$$\begin{aligned} x &= r\cos(\theta) \\ y &= r\sin(\theta) \end{aligned} \tag{1.12}$$

The metric tensor in Cartesian coordinates is $g_{ij} = \delta_{ij}$. Therefore, the square of the arc length is given by:

$$(d\mathbf{s})^2 = \delta_{ij}\frac{\partial v^i}{\partial u^k}\frac{\partial v^j}{\partial u^l}du^k du^l$$

$$= \left(\frac{\partial x}{\partial r}\frac{\partial x}{\partial r} + \frac{\partial y}{\partial r}\frac{\partial y}{\partial r}\right)drdr$$

$$+ \left(\frac{\partial x}{\partial r}\frac{\partial x}{\partial \theta} + \frac{\partial y}{\partial r}\frac{\partial y}{\partial \theta}\right)drd\theta \qquad (1.13)$$

$$+ \left(\frac{\partial x}{\partial \theta}\frac{\partial x}{\partial r} + \frac{\partial y}{\partial \theta}\frac{\partial y}{\partial r}\right)d\theta dr$$

$$+ \left(\frac{\partial x}{\partial \theta}\frac{\partial x}{\partial \theta} + \frac{\partial y}{\partial \theta}\frac{\partial y}{\partial \theta}\right)d\theta d\theta.$$

The partial derivatives are given by:

$$\begin{array}{ll}
\partial x/\partial r = \cos(\theta) & \partial x/\partial \theta = -r\sin(\theta) \\
\partial y/\partial r = \sin(\theta) & \partial y/\partial \theta = r\cos(\theta).
\end{array} \qquad (1.14)$$

With these results, we obtain:

$$(d\mathbf{s})^2 = dr^2 + r^2 d\theta^2. \qquad (1.15)$$

Consequently, the metric tensor in polar coordinates is given in matrix notation by:

$$[g'_{ij}] = \begin{bmatrix} 1 & 0 \\ 0 & r^2 \end{bmatrix}. \qquad (1.16)$$

1.2   MAPPINGS AND SPACES

A mapping (or map) f from a set A, called domain, to another set B, called codomain, denoted by $f : A \rightarrow B$, is a binary relation that assigns elements $a \in A$ (preimage) to elements $b \in B$ (image). For instance, a map $f : X \rightarrow Y$ between two vector spaces X and Y over a field F is linear if it preserves the structure of the vector space, the operation of vector addition, and the operation of scalar multiplication:

$$\forall a_i \in F, \mathbf{x}_i \in X, \; f\left(\sum_i a_i \mathbf{x}_i\right) = \sum_i a_i f(\mathbf{x}_i). \qquad (1.17)$$

A linear map can be represented by a matrix $\mathbf{A} \in \mathbb{R}^{n \times m}$ that maps a vector from $\mathbb{R}^m$ to $\mathbb{R}^n$. This matrix can represent the geometrical operations of rotation, reflection, scaling, and shear, for example.

A metric space (M, d) is a set M equipped with a distance function $d : M \times M \rightarrow \mathbb{R}$, called metric, that satisfies three axioms: i) If two elements $\mathbf{x}, \mathbf{y} \in M$ have the same characteristics, then they are indiscernible ($\mathbf{x} = \mathbf{y}$), and, consequently, the distance between them is zero. Mathematically, this can be stated as $d(\mathbf{x}, \mathbf{y}) = 0 \iff \mathbf{x} = \mathbf{y}$;

ii) The metric should not privilege any of its arguments: $d(\mathbf{x},\mathbf{y}) = d(\mathbf{y},\mathbf{x})$; iii) The distance from going straight from a point $\mathbf{x}$ to another $\mathbf{y}$ must be shorter than taking a detour $\mathbf{z} \in M$ to reach the same endpoint. Therefore, the metric should satisfy the "triangle inequality" (or subadditivity): $d(\mathbf{x},\mathbf{y}) \leq d(\mathbf{x},\mathbf{z}) + d(\mathbf{z},\mathbf{y})$.

Applying the third axiom, we have that $d(\mathbf{x},\mathbf{x}) \leq d(\mathbf{x},\mathbf{y}) + d(\mathbf{y},\mathbf{x})$, but using the second, we see that $d(\mathbf{x},\mathbf{x}) \leq d(\mathbf{x},\mathbf{y}) + d(\mathbf{x},\mathbf{y})$. Now, according to the first axiom, we have that the metric must always be non-negative $d(x,y) \geq 0$. Another important definition, an *ultrametric space* is a metric space where the triangle inequality is replaced by the stricter condition: $d(\mathbf{x},\mathbf{y}) \leq \max\{d(\mathbf{x},\mathbf{z}),d(\mathbf{y},\mathbf{z})\}$.

A related concept, the norm defined over a vector space X and a scalar field F is a real-valued map $X \to \mathbb{R}_+$, denoted by $\|\mathbf{x}\|$ for $\mathbf{x} \in X$, that obeys a collection of properties: i) subadditivity: $\forall \mathbf{x},\mathbf{y} \in X$, $\|\mathbf{x}+\mathbf{y}\| \leq \|\mathbf{x}\| + \|\mathbf{y}\|$, ii) absolute homogeneity: $\forall a \in F$, $\|a\mathbf{x}\| = |a|\|\mathbf{x}\|$, and iii) positive definiteness: $\|\mathbf{x}\| \iff \mathbf{x} = 0$.

A Banach[f] space over a scalar field F is a vector space X equipped with a norm such that the space is complete concerning this norm. This is equivalent to say that Cauchy[g] sequences of $\mathbf{x} \in X$ converge. More formally, for any sequence $\{\mathbf{x}_i\}$, $\mathbf{x}_i \in X$ and any $\forall \varepsilon \in F, \varepsilon > 0$, there is a natural number N such that $\forall i,j \geq N$, $\|\mathbf{x}_i - \mathbf{x}_j\| < \varepsilon$. Therefore, we say that the Banach space is a *complete normed vector space*.

An affine space (A,X) is defined as a set A of points and a vector space X of *displacement vectors*. Points in A cannot be added, but their differences give displacement vectors in X, in other words, it is a vector space without a single origin. Formally it obeys three properties: i) associativity: $\forall a \in A, \mathbf{x},\mathbf{y} \in X$, $a + (\mathbf{x}+\mathbf{y}) = (a+\mathbf{x}) + \mathbf{y}$, ii) right identity: $\forall a \in A, \mathbf{0} \in X$, $a + \mathbf{0} = a$, and iii) subtraction: $\forall a,b \in A, \exists \mathbf{x} \in X$ such that $a = b + \mathbf{x}$.

A map $f : S_1 \to S_2$ between affine spaces $S_1 = (A,X)$ and $S_2 = (B,Y)$ is an affine map if $\forall a \in A, \mathbf{x} \in X$, $f(a+\mathbf{x}) = f(a) + g(\mathbf{x})$, where $g : X \to Y$ is a linear map between vector spaces X and Y. Unlike the linear map, an affine map does not preserve distances and angles but preserves collinearity, and consequently parallelism and ratios of distances. Consequently, an affine map can be represented by $\mathbf{Ax} + \mathbf{b}$, where $\mathbf{A} \in \mathbb{R}^{n \times m}$ is a matrix that maps a vector $\mathbf{x} \in \mathbb{R}^m$ to $\mathbb{R}^n$ and $\mathbf{b} \in \mathbb{R}^n$ is a translation.

1.3 MATRIX CALCULUS

Matrix calculus [6] is a specialized notation that we can use to deal with multidimensional calculus more easily. Let's learn how to do it through some important examples that will be used again in the book.

1.3.1 NOTATION

If $x \in \mathbb{R}$ is a scalar and $\mathbf{y} \in \mathbb{R}^n$ is a vector represented by a column matrix, the derivative of $\mathbf{y}$ to x, is another vector given in index notation by:

$$\left(\frac{\partial \mathbf{y}}{\partial x}\right)_i = \frac{\partial y_i}{\partial x}. \tag{1.18}$$

[f]Stefan Banach (1892–1945) Polish mathematician.
[g]Augustin-Louis Cauchy (1789–1857) French mathematician and physicist.

This can be represented in matrix notation as another column matrix:

$$\frac{\partial \mathbf{y}}{\partial x} = \begin{bmatrix} \partial y_1 / \partial x \\ \partial y_2 / \partial x \\ \vdots \\ \partial y_N / \partial x \end{bmatrix}. \tag{1.19}$$

Inner Product

The *inner product space* is defined as a vector space V over a scalar field $\mathbb{F}$ (typically the real numbers $\mathbb{R}$) equipped with a map $\langle \cdot, \cdot \rangle :$ $V \times V \to \mathbb{F}$ such that for any $x, y, z \in V$ and $a \in \mathbb{F}$, this map has to obey the properties of:

Linearity in the first argument: $\langle s(x+z), y \rangle = s\langle x, y \rangle + s\langle z, y \rangle$,

Hermitian symmetry: $\overline{\langle x, y \rangle} = \langle y, x \rangle$,

Positivity: $\langle x, x \rangle \geq 0$, $\langle x, x \rangle = 0 \iff x = 0$.

When the vector space is $\mathbb{R}^n$ and the scalar field is $\mathbb{R}$, the inner product is known as *scalar product* (or *dot product*). This map $\mathbb{R}^n \times \mathbb{R} \to \mathbb{R}$ is typically denoted by a centered dot instead of the angle brackets. Thus, in index notation, we have for $\mathbf{v}, \mathbf{r} \in \mathbb{R}^n$, that $\mathbf{v} \cdot \mathbf{r} = v_i r^i$. This can be represented by matrices using the transpose: $\mathbf{v}^\top \mathbf{r}$.

We can now find a constant vector $\mathbf{v} \in \mathbb{R}^N$ such that its inner product with $\mathbf{y}$ is x. In matrix notation we can write:

$$x = \mathbf{v}^\top \mathbf{y} \to dx = \mathbf{v}^\top d\mathbf{y}. \tag{1.20}$$

If we write:

$$dx = \frac{dx}{d\mathbf{y}} d\mathbf{y}, \tag{1.21}$$

then we identify:

$$\mathbf{v}^\top = \frac{dx}{d\mathbf{y}} = \begin{bmatrix} dx/dy_1 & dx/dy_2 & \cdots & dx/dy_N \end{bmatrix}. \tag{1.22}$$

Thus, we see that $dx/d\mathbf{y}$ has to be a vector represented by a row matrix. In index notation:

$$\left(\frac{dx}{d\mathbf{y}} \right)_i = \frac{dx}{dy_i}. \tag{1.23}$$

Note that we began with the vector $\mathbf{y}$ represented by a column matrix and obtained dx/dy represented by a row matrix. This is known as *numerator layout* or *Jacobian*[h] *formulation*. Conversely, it would also be possible to begin with a row matrix and obtain a column matrix. This is known as *denominator layout* or *Hessian*[i] *formulation*.

1.3.2 AFFINE MAP

The affine map can be described by the expression $\mathbf{y} = \mathbf{A}\mathbf{x} + \mathbf{b}$, where $\mathbf{y} \in \mathbb{R}^n$ is the output, $\mathbf{A} \in \mathbb{R}^{n \times m}$ represents a linear transformation, $\mathbf{x} \in \mathbb{R}^m$ is the input, and $\mathbf{b} \in \mathbb{R}^m$ represents a translation. In many machine learning applications, particularly those related to neural networks, we need to compute the derivatives of each element. Some other times we need to calculate scalar-valued functions of $\mathbf{y}$.

1.3.2.1 Vector Derivatives

Let's consider the scalar product between a vector $\mathbf{z} \in \mathbb{R}^n$ and a function $f : \mathbb{R} \to \mathbb{R}$ applied to each element of another vector $\mathbf{y} \in \mathbb{R}^n$:

$$a = \mathbf{z}^\top f(\mathbf{y}) = \sum_i^n z_i f(y_i). \tag{1.24}$$

The derivative of this inner product with respect to $\mathbf{y}$ is given by:

$$\begin{aligned} \frac{\partial a}{\partial \mathbf{y}} &= \begin{bmatrix} \frac{\partial a}{\partial y_1} & \frac{\partial a}{\partial y_2} & \cdots & \frac{\partial a}{\partial y_n} \end{bmatrix} \\ &= \begin{bmatrix} z_1 f'(y_1) & z_2 f'(y_2) & \cdots & z_N f'(y_n) \end{bmatrix}. \end{aligned} \tag{1.25}$$

This can be written as[j]:

$$\frac{\partial a}{\partial \mathbf{y}} = \mathbf{z}^\top \circ f'(\mathbf{y})^\top. \tag{1.26}$$

We can also use the *diag* operator to simplify our calculations. This is an *isomorphism* between $\mathbb{R}^n$ and $\mathbb{R}^{n \times n}$. It converts the diagonal of a matrix into a column matrix, and it converts a column matrix into a diagonal matrix.

---

[h]Carl Gustav Jacob Jacobi (1804–1851) German mathematician, adviser of Otto Hesse.

[i]Ludwig Otto Hesse (1811–1874) German mathematician, advised by Carl Gustav Jacob Jacobi, and adviser of Carl Neumann and Gustav Kirchhoff, among others.

[j]The symbol $\circ$ indicates the Hadamard (also known as Schur or element-wise) product between both matrices. Jacques Salomon Hadamard (1865–1963) French mathematician, adviser of Paul Lévy among others. Issai Schur (1875–1941) Russian mathematician advisee of Ferdinand Georg Frobenius.

Isomorphism

Given two mathematical structures[a] $\mathbf{V}$ and $\mathbf{W}$, a bijective map $f : \mathbf{V} \leftrightarrow \mathbf{W}$ is said to be an *isomorphism* between these structures if both are of the same type and there is a reverse map between them. Isomorphic geometric structures that preserve distances between points are also known as *isometry*.

For structures to have the same properties (be similar), it is necessary that the map obeys three properties. Given three mathematical structures A, B, and C: i) any structure has to be similar to itself (reflexivity) $A \cong A$, ii) if $A \cong B$, then $B \cong A$ (symmetry), and iii) if $A \cong B$ and $B \cong C$, then $A \cong C$ (transitivity).

For example, the board of a chess game with black-and-white sites is isomorphic to a chess board with red and green sites since the properties of the board are preserved.

---

[a]A set equipped with some resources such as operators.

With the diag operator, we can write:

$$\frac{\partial a}{\partial \mathbf{y}} = \begin{bmatrix} z_1 & z_2 & \cdots & z_N \end{bmatrix} \begin{bmatrix} f'(y_1) & & & \\ & f'(y_2) & & \\ & & \ddots & \\ & & & f'(y_N) \end{bmatrix} \tag{1.27}$$

$$= \mathbf{z}^\top \operatorname{diag}\left(f'(\mathbf{y})\right).$$

If, for instance, we set $f(\mathbf{y}) = \mathbf{y}$, then this derivative becomes $\partial a/\partial \mathbf{y} = \mathbf{z}^\top$.

1.3.2.2 Jacobian Matrix

Let's consider only the linear map portion of the affine transformation:

$$\mathbf{y} = \mathbf{A}\mathbf{x} = \begin{bmatrix} A_{11} & A_{12} & \cdots & A_{1N} \\ A_{21} & A_{22} & \cdots & A_{2N} \\ \vdots & \vdots & \ddots & \vdots \\ A_{M1} & A_{M2} & \cdots & A_{MN} \end{bmatrix} \begin{bmatrix} x_1 \\ x_2 \\ \vdots \\ x_M \end{bmatrix} \tag{1.28}$$

$$y_m = \sum_{n=1}^{N} W_{mn} x_n.$$

Its derivative to the input $\mathbf{x}$ (*Jacobian matrix*) is given by:

$$\frac{\partial \mathbf{y}}{\partial \mathbf{x}} = \frac{\partial \mathbf{Ax}}{\partial \mathbf{x}} = \begin{bmatrix} \frac{\partial y_1}{\partial \mathbf{x}} \\ \frac{\partial y_2}{\partial \mathbf{x}} \\ \vdots \\ \frac{\partial y_M}{\partial \mathbf{x}} \end{bmatrix} = \begin{bmatrix} \frac{\partial y_1}{\partial x_1} & \frac{\partial y_1}{\partial x_2} & \cdots & \frac{\partial y_1}{\partial x_N} \\ \frac{\partial y_2}{\partial x_1} & \frac{\partial y_2}{\partial x_2} & \cdots & \frac{\partial y_2}{\partial x_N} \\ \vdots & \vdots & \ddots & \vdots \\ \frac{\partial y_M}{\partial x_1} & \frac{\partial y_M}{\partial x_2} & \cdots & \frac{\partial y_M}{\partial x_N} \end{bmatrix}$$

$$= \begin{bmatrix} A_{11} & A_{12} & \cdots & A_{1N} \\ A_{21} & A_{22} & \cdots & A_{2N} \\ \vdots & \vdots & \ddots & \vdots \\ A_{M1} & A_{M2} & \cdots & A_{MN} \end{bmatrix}$$

$$= \mathbf{A}.$$

(1.29)

Consider now three column vectors:

$$\mathbf{r} = \begin{bmatrix} r_1 \\ r_2 \\ \vdots \\ r_N \end{bmatrix}, \quad \mathbf{s} = \begin{bmatrix} s_1 \\ s_2 \\ \vdots \\ s_N \end{bmatrix}, \quad \mathbf{t} = \begin{bmatrix} t_1 \\ t_2 \\ \vdots \\ t_N \end{bmatrix},$$

(1.30)

such that $\mathbf{t}$ is a function of $\mathbf{s}$ that is a function of $\mathbf{r}$. Therefore:

$$\frac{\partial \mathbf{t}}{\partial \mathbf{r}} = \begin{bmatrix} \frac{\partial t_1}{\partial \mathbf{r}} \\ \frac{\partial t_2}{\partial \mathbf{r}} \\ \vdots \\ \frac{\partial t_N}{\partial \mathbf{r}} \end{bmatrix} = \begin{bmatrix} \frac{\partial t_1}{\partial r_1} & \frac{\partial t_1}{\partial r_2} & \cdots & \frac{\partial t_1}{\partial r_N} \\ \frac{\partial t_2}{\partial r_1} & \frac{\partial t_2}{\partial r_2} & \cdots & \frac{\partial t_2}{\partial r_N} \\ \vdots & \vdots & \ddots & \vdots \\ \frac{\partial t_N}{\partial r_1} & \frac{\partial t_N}{\partial r_2} & \cdots & \frac{\partial t_N}{\partial r_N} \end{bmatrix}$$

$$= \begin{bmatrix} \sum_n \frac{\partial t_1}{\partial s_n}\frac{\partial s_n}{\partial r_1} & \sum_n \frac{\partial t_1}{\partial s_n}\frac{\partial s_n}{\partial r_2} & \cdots & \sum_n \frac{\partial t_1}{\partial s_n}\frac{\partial s_n}{\partial r_N} \\ \sum_n \frac{\partial t_2}{\partial s_n}\frac{\partial s_n}{\partial r_1} & \sum_n \frac{\partial t_2}{\partial s_n}\frac{\partial s_n}{\partial r_2} & \cdots & \sum_n \frac{\partial t_2}{\partial s_n}\frac{\partial s_n}{\partial r_N} \\ \vdots & \vdots & \ddots & \vdots \\ \sum_n \frac{\partial t_N}{\partial s_n}\frac{\partial s_n}{\partial r_1} & \sum_n \frac{\partial t_N}{\partial s_n}\frac{\partial s_n}{\partial r_2} & \cdots & \sum_n \frac{\partial t_N}{\partial s_n}\frac{\partial s_n}{\partial r_N} \end{bmatrix}$$

(1.31)

$$= \begin{bmatrix} \frac{\partial t_1}{\partial s_1} & \frac{\partial t_1}{\partial s_2} & \cdots & \frac{\partial t_1}{\partial s_N} \\ \frac{\partial t_2}{\partial s_1} & \frac{\partial t_2}{\partial s_2} & \cdots & \frac{\partial t_2}{\partial s_N} \\ \vdots & \vdots & \ddots & \vdots \\ \frac{\partial t_N}{\partial s_1} & \frac{\partial t_N}{\partial s_2} & \cdots & \frac{\partial t_N}{\partial s_N} \end{bmatrix} \begin{bmatrix} \frac{\partial s_1}{\partial r_1} & \frac{\partial s_1}{\partial r_2} & \cdots & \frac{\partial s_1}{\partial r_N} \\ \frac{\partial s_2}{\partial r_1} & \frac{\partial s_2}{\partial r_2} & \cdots & \frac{\partial s_2}{\partial r_N} \\ \vdots & \vdots & \ddots & \vdots \\ \frac{\partial s_N}{\partial r_1} & \frac{\partial s_N}{\partial r_2} & \cdots & \frac{\partial s_N}{\partial r_N} \end{bmatrix}$$

$$= \frac{\partial \mathbf{t}}{\partial \mathbf{s}} \frac{\partial \mathbf{s}}{\partial \mathbf{r}}.$$

Therefore, the chain rule applies in this case. Also, note that this is an operation of *contraction*:

$$\left(\frac{\partial \mathbf{t}}{\partial \mathbf{r}} \right)_{ij} = \sum_n \left(\frac{\partial \mathbf{t}}{\partial \mathbf{s}} \right)_{in} \left(\frac{\partial \mathbf{s}}{\partial \mathbf{r}} \right)_{nj}.$$

(1.32)

This operation is illustrated in Fig. 1.2.

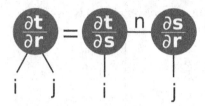

Figure 1.2   Representation of the chain rule using contraction

1.3.2.3   Matrix Derivatives

We now want to compute the derivative of $\mathbf{y} \in \mathbb{R}^n$ to the linear transformation matrix $\mathbf{A} \in \mathbb{R}^{n \times m}$. This results in a third-order tank tensor:

$$\frac{\partial \mathbf{y}}{\partial \mathbf{A}} = \begin{bmatrix} \frac{\partial y_1}{\partial \mathbf{A}} \\ \vdots \\ \frac{\partial y_n}{\partial \mathbf{A}} \end{bmatrix} = \begin{bmatrix} \frac{\partial y_1}{\partial A_{1,*}} & \frac{\partial y_1}{\partial A_{2,*}} & \cdots & \frac{\partial y_1}{\partial A_{m,*}} \\ \frac{\partial y_2}{\partial A_{1,*}} & \frac{\partial y_2}{\partial A_{2,*}} & \cdots & \frac{\partial y_2}{\partial A_{m,*}} \\ \vdots & \vdots & \ddots & \vdots \\ \frac{\partial y_n}{\partial A_{1,*}} & \frac{\partial y_n}{\partial A_{2,*}} & \cdots & \frac{\partial y_n}{\partial A_{m,*}} \end{bmatrix} = \tag{1.33}$$

$$=$$

In this case, the index tensor notation becomes more convenient:

$$T_{ijk} = \frac{\partial y_i}{\partial A_{jk}}. \tag{1.34}$$

For our affine map, we have:

$$y_i = \sum_m A_{im} x_m + b_i,$$
$$\frac{\partial y_i}{\partial A_{jk}} = \sum_m x_m \delta(i-j)\delta(m-k), \tag{1.35}$$
$$T_{ijk} = x_k \delta(i-j).$$

If this tensor appears inside the chain rule, it is necessary to do a contraction. For example, if z is a function of $\mathbf{y}$, we get:

$$\frac{\partial z}{\partial \mathbf{A}} = \frac{\partial z}{\partial \mathbf{y}} \frac{\partial \mathbf{y}}{\partial \mathbf{A}} = \frac{\partial z}{\partial \mathbf{y}} \mathbf{T}$$

$$\frac{\partial z}{\partial A_{jk}} = \sum_{i=1} \frac{\partial z}{\partial y_i} \frac{\partial y_i}{\partial W_{jk}}$$

$$= \sum_{i=1} \frac{\partial z}{\partial y_i} x_k \delta(i-j) \qquad (1.36)$$

$$= x_k \frac{\partial z}{\partial y_j}$$

$$\frac{\partial z}{\partial \mathbf{A}} = \mathbf{x} \frac{\partial z}{\partial \mathbf{y}}.$$

This operation is illustrated in Fig. 1.3.

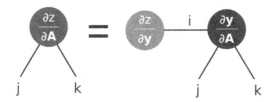

Figure 1.3   Chain rule involving a vector and a tensor of rank three

Another possibility is to do a *tensor vectorization*:

$$\mathbf{Ax} = \begin{bmatrix} A_{11} & A_{12} \\ A_{21} & A_{22} \\ A_{31} & A_{32} \end{bmatrix} \begin{bmatrix} x_1 \\ x_2 \end{bmatrix} = \begin{bmatrix} A_{11}x_1 + A_{12}x_2 \\ A_{21}x_1 + A_{22}x_2 \\ A_{31}x_1 + A_{32}x_2 \end{bmatrix}$$

$$= \begin{bmatrix} x_1 & 0 & 0 & x_2 & 0 & 0 \\ 0 & x_1 & 0 & 0 & x_2 & 0 \\ 0 & 0 & x_1 & 0 & 0 & x_2 \end{bmatrix} \begin{bmatrix} A_{11} \\ A_{21} \\ A_{31} \\ A_{12} \\ A_{22} \\ A_{32} \end{bmatrix} \qquad (1.37)$$

$$= \left[\begin{bmatrix} x_1 & 0 & 0 \\ 0 & x_1 & 0 \\ 0 & 0 & x_1 \end{bmatrix} \begin{bmatrix} x_2 & 0 & 0 \\ 0 & x_2 & 0 \\ 0 & 0 & x_2 \end{bmatrix} \right] \mathrm{vec}(\mathbf{A})$$

$$= \begin{bmatrix} x_1 \mathbb{I} & x_2 \mathbb{I} \end{bmatrix} \mathbf{a}$$

$$= (\mathbf{x}^\mathsf{T} \otimes \mathbb{I}) \mathbf{a}.$$

Therefore,

$$\frac{d}{d\mathbf{A}}(\mathbf{Ax}) \mapsto \frac{d}{d\mathbf{a}}(\mathbf{Ax}) = \mathbf{x}^\mathsf{T} \otimes \mathbb{I}. \qquad (1.38)$$

In this procedure, the isomorphism $\text{vec}(\mathbf{A}) : \mathbb{R}^{M \times N} \to \mathbb{R}^M \otimes \mathbb{R}^N = \mathbb{R}^{MN}$ is known as the vectorization of matrix $\mathbf{A}$. In Julia, vectorization is performed using:

```
M = [1 2;3 4]
v = vec(M)
```

The inverse operation $\text{vec}^{-1}(\mathbf{a}) : \mathbb{R}^{MN} \to \mathbb{R}^{M \times N}$ is known as *matricization* or *reshaping*. In Julia:

```
v = [1,2,3,4]
M = reshape(v,2,2)
```

It is simple to show that also using this technique we obtain:

$$\frac{\partial z}{\partial \mathbf{y}} \frac{\partial \mathbf{y}}{\partial \mathbf{A}} = \mathbf{x} \frac{\partial z}{\partial \mathbf{y}}.$$

The operation $\mathbf{A} \otimes \mathbf{B}$ is the Kronecker[k] product between matrices $\mathbf{A} \in \mathbb{R}^{M \times N}$ and $\mathbf{B} \in \mathbb{R}^{P \times Q}$ defined as:

$$\mathbf{A} \otimes \mathbf{B} = \begin{bmatrix} a_{11}\mathbf{B} & \dots & a_{1N}\mathbf{B} \\ \vdots & \ddots & \vdots \\ a_{M1}\mathbf{B} & \dots & a_{MN}\mathbf{B} \end{bmatrix}. \tag{1.39}$$

For example, if $\mathbf{x} = \begin{bmatrix} x_1 & x_2 \end{bmatrix}$ and $\mathbb{I} = \begin{bmatrix} 1 & 0 \\ 0 & 1 \end{bmatrix}$, their Kronecker product is:

$$\begin{aligned} \mathbf{x} \otimes \mathbb{I} &= \begin{bmatrix} x_1\mathbf{y} & x_2\mathbf{y} \end{bmatrix} \\ &= \begin{bmatrix} x_1 \begin{bmatrix} 1 & 0 \\ 0 & 1 \end{bmatrix} & x_2 \begin{bmatrix} 1 & 0 \\ 0 & 1 \end{bmatrix} \end{bmatrix} \\ &= \begin{bmatrix} x_1 & 0 & x_2 & 0 \\ 0 & x_1 & 0 & x_2 \end{bmatrix}. \end{aligned} \tag{1.40}$$

The Kronecker product in Julia is obtained using:

```
A = [1 2;3 4]
B = [5 6;7 8]
C = kron(A,B)
```

---

[k]Leopold Kronecker (1823–1891) German mathematician, advised by Peter Gustav Lejeune Dirichlet.

2 Probability Theory

Often in machine learning, one needs to work with uncertain or probabilistic data. Probability theory offers the tools to represent and manipulate models that make predictions and decisions based on random and incomplete information[a]. This chapter introduces the necessary background in probability theory and statistical inference that underlie these probabilistic models. We will begin with the concept of probability space, which is fundamental to the comprehension of probability distribution and statistical reasoning. Next, we will discuss random variables, which are essential for representing and analyzing data in machine learning. This is complemented with Bayes methods, which are widely used for parameter estimation and decision making. As the chapter progresses, we will discuss the amount of information that is contained in random variables and how to use it to establish relationships between distributions. Also, we will see how to estimate the parameters of distributions using the technique of maximum likelihood and kernel density estimation. Finally, we will end this chapter discussing Markov chains, a memoryless stochastic process used in many applications, including, for example, speech recognition.

2.1 PROBABILITY SPACE

Consider a process in which the outcomes are uncertain. The mathematical framework that captures such random situations is called a *probability space*. It consists of three components: the sample space, denoted by the symbol Ω, which is a set of all possible outcomes that may occur in the process; the event space (a σ-algebra), denoted by the symbol Σ, which is a collection of subsets of Ω representing all possible events that can occur; and the probability measure, denoted by the symbol P, which assigns a probability between 0 and 1 to each event in the event space, representing the likelihood of such an event occurring. In other words, a probability space is a 3-tuple (Ω, Σ, P) that enables us to mathematically model and analyze uncertain events.

To illustrate the concept of a probability space, let's consider the example of flipping a coin. In this stochastic process, the sample space, denoted by the symbol Ω, consists of the possible outcomes, which are Head and Tail, represented as $\Omega = \{H, T\}$.

To properly model the probability of any event occurring, we must consider all possible events, including the possibility of no event occurring. This means that the event space, denoted by the symbol Σ, must contain both the empty set and the sample space itself, i.e., $\Sigma = \{\emptyset, \Omega\}$. This is, consequently, the smallest possible σ-algebra.

Furthermore, if we include one event in the event space, we must also include its complement. For example, if we include the event H, we must also include the event

---

[a]Good references for this chapter are [7–9].

DOI: 10.1201/9781003350101-2

T, because if we knew that H occurred, then we would immediately know that T did not occur. Consequently, the biggest σ-algebra is $\Sigma = \{\emptyset, \Omega, \{H\}, \{T\}\}$, which is the power set of the sample space.

In summary, the event space must always contain the empty set, the sample space, and subsets of Ω that are closed under the complement and countable operations of union and intersection.

If we assume that the coin used in the example is unbiased, then the probability of obtaining either a Head or a Tail by flipping it should be equal. This means that the probability function, denoted by the symbol P, should assign equal probabilities to both outcomes, i.e., $P(H) = P(T) = 0.5$.

Moreover, since the sample space contains all possible outcomes, the probability of any event that can occur must be equal to the sum of probabilities of the individual outcomes that make up the event. In other words, the probability of the entire sample space Ω must be equal to 1 since this represents the probability that one of the outcomes in the sample space will occur. Therefore, $P(\Omega) = 1$.

Using these probabilities and the event space we previously defined, we can assign probabilities to other events. For example, the probability of getting exactly one Head in two flips can be calculated by considering the event $\{HT, TH\}$, which contains two outcomes with exactly one Head. Since each outcome in this event has a probability $0.25 = 0.5 \times 0.5$, the probability of the event itself is 0.5.

In the context of the coin flipping example we have been discussing, the pair of mathematical objects consisting of the sample space Ω and the event space Σ, denoted as (Ω, Σ), is known as a *measurable space* or a *Borel space*. This pair defines the basic structure required to assign probabilities to events.

When we introduce the probability measure P into the measurable space, creating the 3-tuple (Ω, Σ, P), we obtain a *measure space*. This allows us to assign probabilities to the events defined by the event space. Therefore, a measure space extends the basic structure of a measurable space by incorporating a probability measure, allowing us to quantify the likelihood of different events occurring.

It is important to note that the probability measure P is a mathematical function that satisfies certain properties, making it a *measure*. Specifically, P is a non-negative function, meaning that it only takes non-negative values for all events. Additionally, $P(\emptyset) = 0$, which is consistent with the idea that the probability of an impossible event happening is zero. Finally, $P(\bigcup_k E_k) = \sum_k P(E_k)$, $\forall E_k \in \Sigma$, meaning that the probability of a union of events is equal to the sum of their probabilities.

2.1.1 RANDOM VARIABLES

This framework of a measure space allows us to more formally define a *random variable*. A random variable is a function $X : \Omega \to \Omega'$ that maps the sample space Ω to another measurable space Ω', such that events in Ω' can be assigned probabilities. This is done by ensuring that the inverse image of every measurable set in Ω' under the random variable is a measurable set in Ω: $\{\omega : X(\omega) \in B\} \in \Sigma$, $\forall B \in \Sigma'$. In other words, a random variable is a measurable function in a Borel space (Ω', Σ').

In the context of flipping a coin, a simple example of a random variable could be the function X defined as $X(H) = 1$, $X(T) = 0$. In other words, X maps the outcomes of the coin flip to either 1 or 0, depending on whether the outcome is a head or a tail, respectively. Therefore, this random variable measures the number of heads obtained in a single coin flip.

Now we can ask, for example, what is the probability that we get one head in a single flip. This event can be written as $X^{-1}(1) = H$, which is a subset of the event space Σ. Since the coin is unbiased, the probability of getting one head in a single flip is $P(X^{-1}(1)) = P(H) = 0.5$.

2.1.2   CONDITIONAL PROBABILITY

By defining a random variable, we can now use the tools of probability theory to analyze and predict the behavior of the underlying stochastic process. To explore it further, let's consider the case of drawing two balls from a bag without replacement and, we know that there are two red balls and two blue balls in the bag. The probability of drawing either a blue or a red ball from the bag is 0.5. However, the probability of drawing a second ball depends on the information obtained in the first draw. For example, let's consider that the first drawn ball was red. Therefore, the probability of the second drawn ball being blue is 2 out of 3.

This can be expressed mathematically as the *conditional probability* of event A given B, denoted by $P(A|B)$, which can be calculated by dividing the probability of A and B occurring together by the probability of B occurring alone. In the example, A represents drawing a blue ball on the second draw, whereas B represents drawing a red ball on the first draw. The probability of drawing a red ball on the first draw is 2 out of 4 balls. The number of ordered pairs is $2\binom{4}{2} = 12$, corresponding to:

$$\{Red_1, Red_2\} \quad \{Red_1, Blue_1\} \quad \{Red_1, Blue_2\}$$
$$\{Red_2, Red_1\} \quad \{Red_2, Blue_1\} \quad \{Red_2, Blue_2\}$$
$$\{Blue_1, Red_1\} \quad \{Blue_1, Red_2\} \quad \{Blue_1, Blue_2\}$$
$$\{Blue_2, Red_1\} \quad \{Blue_2, Red_2\} \quad \{Blue_2, Blue_1\}.$$

From these subsets, four of them correspond to drawing together a red ball on the first draw and a blue ball on the second draw:

$$\{Red_1, Blue_1\} \quad \{Red_1, Blue_2\}$$
$$\{Red_2, Blue_1\} \quad \{Red_2, Blue_2\}.$$

Consequently, $P(A \cap B) = 4/12$, and:

$$P(A|B) = \frac{P(A \cap B)}{P(B)} = \frac{1/3}{1/2} = \frac{2}{3}.$$

Observe that if the chance of both events A and B happening together is just the product of the probabilities of A and B, then we say that these events are independent.

2.1.2.1　Baye's Theorem

Let's switch our attention now to a similar problem involving two bags X and Y. Bag X contains 3 red balls and 2 blue balls, whereas bag Y contains 1 red ball and 4 blue balls. If you pick one bag at random with an equal chance, what is the probability that you chose bag X given that a red ball was drawn? To solve this, let's consider A the event of choosing bag X and B the event of drawing a red ball. Consequently, we are interested in finding:

$$P(A|B) = \frac{P(A \cap B)}{P(B)}.$$

Conversely, we can invert this problem and ask what is the probability of drawing a red ball given that bag X was chosen. For this, we have:

$$P(B|A) = \frac{P(B \cap A)}{P(A)}.$$

Combining both equations, we find:

$$P(A|B)P(B) = P(B|A)P(A) \tag{2.1}$$

Thomas Bayes[b] was the first to derive this result, which is now commonly known as *Bayes' Theorem* [7]. Let's apply it to our problem by writing:

$$P(A|B) = \frac{P(B|A)P(A)}{P(B)}. \tag{2.2}$$

Written in this form, A is known as a *hypothesis*, $P(A)$ is the *prior probability* that estimates the probability of the hypothesis before the evidence B is observed, $P(A|B)$ is the *posterior probability*, or the probability of the hypothesis given the evidence, $P(B|A)$ is the *likelihood* of observing the evidence given the hypothesis, and $P(B)$ is the *marginal likelihood*, which corresponds to the probability of generating an observed sample from a prior.

In the example, the prior probability $P(A)$ of choosing bag X is $1/2$. The likelihood $P(B|A)$ of observing a red ball given the hypothesis of choosing bag X is three balls out of five. The marginal likelihood $P(B)$ of generating an observed sample (drawing a red ball) from the prior can be computed using the *law of total probability*, which states that the probability of an event B can be calculated by summing the probabilities of drawing a red ball under each bag. Mathematically, the latter is given by:

$$\begin{aligned} P(B) &= P(B,A) + P(B,\bar{A}) \\ &= P(B|A)P(A) + P(B|\bar{A})P(\bar{A}) \\ &= \frac{3}{5} \times \frac{1}{2} + \frac{1}{5} \times \frac{1}{2} = \frac{2}{5}. \end{aligned} \tag{2.3}$$

---

[b]Thomas Bayes (1701–1761) English statistician and philosopher.

Consequently, the probability of having chosen bag X given that a red ball was drawn is given by:

$$P(A|B) = \frac{P(B|A)P(A)}{P(B)}$$
$$= \frac{\frac{3}{5} \times \frac{1}{2}}{\frac{2}{5}} = \frac{3}{4}. \tag{2.4}$$

Another important result from Bayes' theorem is the chain rule. Given a set of finitely many events $\{X_1, \ldots, X_N\}$, one can write:

$$P(X_1, \ldots, X_N) = P(X_N | X_1, \ldots, X_{N-1}) P(X_1, \ldots, X_{N-1}). \tag{2.5}$$

By induction, we get:

$$P(X_1, \ldots, X_N) = P(X_N | X_1, \ldots, X_{N-1}) P(X_{N-1} | X_1, \ldots, X_{N-2}) \ldots P(X_2 | X_1) P(X_1)$$
$$= \prod_{n=1}^{N} P(X_n | X_1, \ldots, X_{n-1}). \tag{2.6}$$

2.1.3 PROBABILITY DISTRIBUTION

Instead of assigning a single probability value to an event, a probability distribution function can be used to define a distribution of probability values. For a random variable X defined on a probability space (Ω, Σ, P), its cumulative distribution function (CDF) maps any value $x \in \mathbb{R}$ to the probability that X takes on a value less than or equal to x. This is mathematically defined as:

$$F_X(x) := P[X \leq x] = P(\{\omega \in \Omega | X(\omega) \leq x\}). \tag{2.7}$$

To advance, we need the concept of the Lebesgue[c] measure. Given a measurable space (Ω, Σ), the Lebesgue measure is a measure $\mu : \Sigma \to [0, \infty]$ equipped with the property of translation-invariance, meaning that the measure of a set is the same regardless of its position on the real line. For example, the Lebesgue measure over the interval (a, b) is $\mu((a, b)) = b - a$.

Based on this measure, a Lebesgue integral can be created. For this, we decompose a positive simple function in a linear combination of indicator functions[d] over a Borel[e] set B:

$$g(x) = \sum_i c_i \mathbf{1}_{A_i}(x), \quad \bigcup_j A_j = B. \tag{2.8}$$

---

[c]Henri Léon Lebesgue (1875–1941) French mathematician advised by Émile Borel.

[d]For a subset A of a set X, the indicator function is defined as $\mathbf{1}_A(x) := \begin{cases} 1 & \text{if } x \in A \\ 0 & \text{otherwise.} \end{cases}$

[e]Félix Édouard Justin Émile Borel (1871–1956) French mathematician. Borel was the adviser of Henri Lebesgue, among others. A Borel set is a set that can be constructed from open sets using operations of union, intersection, and complement.

The Lebesgue integral is then defined as:

$$\int_B g(x)\mu(dx) = inf\left\{\sum_i c_i\mu(A_i)\right\}, \tag{2.9}$$

where inf is the *infimum* defined as the biggest element of an ordered set that is smaller or equal to all elements of a subset for which it is being calculated.

Now, we can still integrate a function $f(x)$ over a Borel set by expressing the integral over the entire $\mathbb{R}$ domain using an indicator function. Furthermore, the restriction imposed by the indicator function can be carried to the metric itself:

$$\int_\mathbb{R} g(x)\mathbf{1}_B\mu(dx) = \int_\mathbb{R} fd\lambda. \tag{2.10}$$

Inspecting this last equation, it is easy to observe that $d\lambda = fd\mu$, where $f = \mathbf{1}_B$ is the density of λ, known as the *Radon-Nikodym derivative*[f] [10] with respect to the Lebesgue measure [11].

This framework can be used with the CDF considering that the probability of anything happening must be 1:

$$\int_\Omega dF_X(x) = \int f(x)dx = 1, \tag{2.11}$$

where $f(x)$, if it exists, is the *probability density function* (PDF) defined as:

$$dF_X(x) = f(x)dx \to f(x) = \frac{dF_X(x)}{dx}. \tag{2.12}$$

When the random variable is discrete, the density is called the *probability mass function* (PMF).

Example 2.1.1. Gaussian Distribution

Consider the case of throwing darts into a board. Although the darts are aimed at the position $(0,0)$, they end up in a random position (X,Y). The density of positions is invariant over rotation so that only the distance from the center is important. Moreover, the random variables X and Y should be independent. Therefore, the joint density $f_{X,Y}(x,y)$ should be given by a function $g(x^2+y^2)$. Since the distributions are independent, we have $f_{X,Y}(x,y) = f_{X,Y}(x,0)f_{X,Y}(0,y)$. Consequently, we have:

$$g(x^2+y^2) = g(r^2) = g(x^2)g(y^2),$$

which implies that the function g must be exponential: $g(r) = Ae^{-Br}$ and $f_X(x) = g(x^2) = Ae^{-Bx^2}$. If the mean of the distribution is μ, then the density becomes $f_X(x) = Ae^{-B(x-\mu)^2}$.

---

[f]Johann Karl August Radon (1887–1956) Austrian mathematician, and Otto Marcin Nikodym (1887–1974) Polish mathematician, advised by Waclaw Sierpiński and Marian Smoluchowski.

By definition, the integral of the density over the whole domain must be 1. Therefore:

$$A \int_{-\infty}^{\infty} e^{-B(x-\mu)^2} dx = 1 \therefore A = \sqrt{\frac{B}{\pi}}. \tag{2.13}$$

The second central moment of a distribution is called *variance*:

$$\sigma^2 = \sqrt{\frac{B}{\pi}} \int_{-\infty}^{\infty} x^2 e^{-B(x-\mu)^2} dx = \frac{1}{2B}. \tag{2.14}$$

With this, we arrive at the *Gaussian*[g] or *normal* distribution:

$$f_X(x) = \frac{1}{\sqrt{2\pi\sigma^2}} \exp\left(-\frac{(x-\mu)^2}{2\sigma^2}\right). \tag{2.15}$$

When a random variable X is normally distributed with mean μ and variance σ^2, it can be denoted by $X \sim \mathcal{N}(\mu, \sigma)$. ∎

Example 2.1.2. Kolmogorov-Smirnof Test

The Kolmogorov[h]-Smirnov[i] (KS) test [12] is a non-parametric test used to determine whether a sample was taken from a population with a specific distribution F.

Consider an empirical distribution F_n for n independent and identically distributed (i.i.d.) observations obtained with:

$$F_n(x) = \frac{1}{N} \sum_{m}^{n} I_x(X_m), \tag{2.16}$$

where I_x is the indicator function that gives 1 if $X_m \leq x$ and 0 otherwise.

The KS test is given by:

$$D_n = \sup_x |F_n(x) - F(x)|, \tag{2.17}$$

where $\sup_x$ is the *supremum* (the smallest element that is greater or equal to every element in a set) of the set.

The hypothesis that the sample is drawn from the reference distribution is rejected with a confidence level α if:

$$\sqrt{n}D_n > K_\alpha, \tag{2.18}$$

where K_α is given by:

$$P(K \leq K_\alpha) = 1 - \alpha, \tag{2.19}$$

---

[g]Johann Carl Friedrich Gauss (1777–1855) German mathematician and physicist, adviser of Bernhard Riemann, among others. Gauss was awarded the Copley Medal in 1838.
[h]Andrey Nikolaevich Kolmogorov (1903–1987) Russian mathematician.
[i]Nikolai Vasilyevich Smirnov (1900-1966) Russian mathematician.

where K is the Kolmogorov distribution:

$$P(K \leq x) = 1 - 2 \sum_{k=1}^{\infty} (-1)^{k-1} e^{-2(kx)^2}. \tag{2.20}$$

For $n > 35$, the asymptotic values are approximated by:

$$K_\alpha \approx \sqrt{\frac{-1/2 \ln\left(\frac{\alpha}{2}\right)}{n}}. \tag{2.21}$$

■

2.2 INFORMATION

What kind of information[j] is contained in a random variable? If a random variable has a probability measure of 100%, then we know the outcome with certainty, and therefore the information contained in the variable is zero. On the other hand, the less probable an event is, the more surprising it is and the more information it carries. Hence, the information contained in a random variable is related to its probability measure, and there is no reason to expect sudden jumps in information as the probability measure is varied. Therefore, we can conclude that information is a monotonically decreasing function of probability.

If two events are independent, then the occurrence of one event does not affect the probability of the other event occurring. Therefore, the amount of information gained from knowing that both events occurred is simply the sum of the information gained from knowing that each event occurred separately. Consequently, we have:

$$I_{X,Y}(p_X(x)p_Y(y)) = I_X(p(x)) + I_Y(p(y)). \tag{2.22}$$

One of the simplest functions that satisfy these requirements is $I_X(x) := -\log(p_X(x))$. Therefore, if the probability is 1, the information is 0, whereas if the probability is 0, the information tends to infinity, which corresponds to total surprise.

The expected information content of a random variable is then given by:

$$\begin{aligned} H(X) = \langle I_X \rangle &= \sum_x p_X(x) I_X(x) \\ &= -\sum_x p_X(x) \log(p_X(x)). \end{aligned} \tag{2.23}$$

Being conceived by Shannon[k], this expression is called *Shannon entropy* of the random variable.

Take for example the tossing of a coin expressed by a random variable X that takes 0 for tails and 1 for heads with equal probability. The entropy of this random variable is:

$$\begin{aligned} H(X) &= -\sum_{x \in 0,1} P(X = x) \log_2 P(X = x) \\ &= -\left[\frac{1}{2} \log_2 \left(\frac{1}{2}\right) + \frac{1}{2} \log_2 \left(\frac{1}{2}\right) \right] = 1. \end{aligned} \tag{2.24}$$

---

[j]A good resource on information theory can be found in [13].

[k]Claude Elwood Shannon (1916–2001) American mathematician, electrical engineer, and computer scientist. Shannon was awarded many prizes including the IEEE Medal of Honor in 1966.

Observe that the logarithm was taken with base 2. This implies that the entropy of the coin toss is 1 bit, which is the amount of information required to represent the outcome of a fair coin toss.

2.2.1 STATISTICAL DIVERGENCE

The statistical divergence[1] can be considered the difference in entropy between two probability distributions. Given two probability distributions P and Q over the same sample space Ω, we can compare the amount of information contained in these distributions by comparing their entropies.

If the two distributions are very similar, we would expect their entropies to be close to each other. Conversely, if the two distributions are very different, we would expect their entropies to be farther apart. The measure of this difference is called the *Kullback-Leibler*[m] (KL) divergence [14], denoted by $D_{KL}(P|Q)$ and defined as:

$$D_{KL}(P|Q) = \sum_{x \in \mathcal{X}} P(x) \log \left(\frac{P(x)}{Q(x)} \right)$$
$$= \langle \log(P) \rangle_P - \langle \log(Q) \rangle_P . \tag{2.25}$$

Intuitively, this formula measures the average amount of extra information needed to encode a message generated from P when we use a code optimized for Q and is sometimes referred to as the "information gain" or "information loss" when moving from Q to P.

Example 2.2.1. Divergence Between Normally-Distributed Random Variables

For example, given two random variables $P \sim \mathcal{N}(\mu_1, \sigma_1)$ and $Q \sim \mathcal{N}(\mu_2, \sigma_2)$, the KL divergence between them is:

$$D_{KL}(P|Q) = \langle \log(p(x)) - \log(q(x)) \rangle_P$$
$$= \left\langle \log \left(\frac{\sigma_2}{\sigma_1} \right) + \frac{1}{2} \left[\left(\frac{x - \mu_2}{\sigma_2} \right)^2 - \left(\frac{x - \mu_1}{\sigma_1} \right)^2 \right] \right\rangle_P \tag{2.26}$$
$$= \log \left(\frac{\sigma_2}{\sigma_1} \right) + \frac{\sigma_1^2 + (\mu_1 - \mu_2)^2}{2\sigma_2^2} - \frac{1}{2} .$$

More generally, the KL divergence is one type of *f-divergence* given by:

$$D_f(P \| Q) = \int f \left(\frac{dP}{dQ} \right) dQ, \tag{2.27}$$

where $f : [0, \infty] \to (-\infty, \infty]$ is a convex function, called the *generator* of the divergence.

---

[1] Not to be confused with the standard mathematical divergence.
[m] Solomon Kullback (1907–1994) and Richard Leibler (1914–2003) American mathematicians.

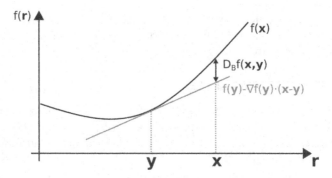

Figure 2.1 Illustration of the Bregman divergence

Moreover, KL divergence is a type of *Bregman*[n] divergence. The Bregman divergence between two points $\mathbf{x}$ and $\mathbf{y}$ in a convex space Ω with respect to a strictly convex function $f : \Omega \to \mathbb{R}$ is defined as the difference between the value evaluated at point $\mathbf{x}$ and its first-order Taylor expansion around $\mathbf{y}$ evaluated at $\mathbf{x}$:

$$D_B f(\mathbf{x}, \mathbf{y}) = f(\mathbf{x}) - f(\mathbf{y}) - \nabla f(\mathbf{y}) \cdot (\mathbf{x} - \mathbf{y}). \qquad (2.28)$$

This is illustrated in Fig. 2.1. Bregman divergence can be used, for example, to find the common point in a set of convex sets [15].

To see how the Kullback-Leibler divergence is a Bregman divergence, take $f(\mathbf{x}) = \sum_i x_i \log(x_i)$ (negative entropy), where x is a probability distribution. Calculating the Bregman divergence gives:

$$
\begin{aligned}
D_B f(\mathbf{p}, \mathbf{q}) &= \mathbf{p} \cdot \log(\mathbf{p}) - \mathbf{q} \cdot \log(\mathbf{q}) - (\log(\mathbf{q}) + \mathbf{e}) \cdot (\mathbf{p} - \mathbf{q}) \\
&= \mathbf{p} \cdot \log(\mathbf{p}) - \mathbf{q} \cdot \log(\mathbf{q}) - \mathbf{p} \cdot \mathbf{e} + \mathbf{q} \cdot \mathbf{e} \\
&= \sum_i p_i \log\left(\frac{p_i}{q_i}\right) = D_{KL}(P|Q).
\end{aligned}
\qquad (2.29)
$$

Observe that, if we take $f(\mathbf{x}) = |\mathbf{x}|^2$, then the Bregman divergence returns the Euclidean distance between $\mathbf{x}$ and $\mathbf{y}$. In fact, both f-divergences and the Bregman divergence are non-negative and linear, but neither satisfies the triangle inequality nor are in general symmetric. Consequently, f-divergences and Bregman divergence are not metrics (see Sec. 1.2.), but since they resemble metrics, they are often called *statistical distances* and are used to quantify the distance between statistical objects such as random variables and distributions.

λ-divergences symmetrize the Kullback-Leibler divergence as [16]:

$$D_\lambda(P|Q) = \lambda D_{KL}(P|\lambda P + (1-\lambda)Q) + (1-\lambda)D_{KL}(Q|\lambda P + (1-\lambda)Q). \quad (2.30)$$

---

[n]Lev Meerovich Bregman (1941–2023) Russian/Israeli mathematician.

When $\lambda = 1/2$, the divergence obeys the triangle inequality, can be regarded as a true distance, and is referred to as *Jensen-Shannon* (JS) divergence [17]:

$$D_{JS} = \frac{1}{2}\left[D_{KL}(P|M) + D_{KL}(Q|M)\right], \tag{2.31}$$

where $M = 1/2(P+Q)$.

2.2.1.1 Fisher Information

The Kulback-Leibler divergence holds some connection with information metrics. For example, consider that distribution $Q = P_\theta$ has some parameter set θ and distribution P_{θ_o} is very close to it. Taylor expanding the KL divergence around θ_o, we get:

$$D_{KL}(P_{\theta_o}\|P_\theta) \approx D_{KL}(P_{\theta_o}\|P_{\theta_o}) + \nabla_\theta^\top D_{KL}(P_{\theta_o}\|P_\theta)\Big|_{\theta=\theta_o}(\theta - \theta_o)+ $$
$$+\frac{1}{2}(\theta - \theta_o)^\top \mathbf{H}_D(\theta - \theta_o), \tag{2.32}$$

where $\mathbf{H}_D$ is the Hessian of the KL divergence.

Both $D_{KL}(P_{\theta_o}\|P_{\theta_o})$ and the gradient of the KL divergence are zero. From Eq. 2.25:

$$\nabla_\theta^\top D_{KL}(P_{\theta_o}\|P_\theta)\Big|_{\theta=\theta_o} = -\left\langle \frac{1}{P}\frac{\partial P}{\partial \theta_i} \right\rangle_P = -\sum_{x\in\chi}\frac{\partial P}{\partial \theta_i} = -\frac{\partial}{\partial \theta_i}\sum_{x\in\chi}P = 0. \tag{2.33}$$

The Hessian is given by:

$$\frac{\partial^2}{\partial \theta_j \partial \theta_i}D_{KL}(P_{\theta_o}\|P_\theta) = -\left\langle \frac{\partial}{\partial \theta_j}\left(\frac{1}{P}\frac{\partial P}{\partial \theta_i}\right) \right\rangle_P = -\left\langle -\frac{1}{P^2}\frac{\partial P}{\partial \theta_j}\frac{\partial P}{\partial \theta_i} + \frac{1}{P}\frac{\partial^2 P}{\partial \theta_j \partial \theta_i} \right\rangle_P$$
$$g_{ji} = \left\langle \frac{\partial}{\partial \theta_j}\log(P)\frac{\partial}{\partial \theta_i}\log(P) \right\rangle$$
$$\mathbf{H}_D = \left\langle (\nabla_\theta \log P_\theta)(\nabla_\theta \log P_\theta)^\top \right\rangle_{P,\theta=\theta_o}. \tag{2.34}$$

This is exactly the Fisher[o] information metric [18]. Therefore:

$$D_{KL}(P_{\theta_o}\|P_\theta) \approx \frac{1}{2}(\theta - \theta_o)^\top \mathbf{H}_D(\theta_o)(\theta - \theta_o). \tag{2.35}$$

Thus, the Fisher metric tells us how much the distribution changes if the parameters are differentially updated in a certain direction in parameter space.

---

[o]Ronald Aylmer Fisher (1890–1962) British mathematician, winner of many prizes including the Copley Medal in 1955.

2.2.2 MAXIMUM LIKELIHOOD

Given N observations $x_1, \ldots, x_N$ from i.i.d. random variables, the likelihood [19] function gives the probability of their occurrence given a probability density $f(x|\theta)$ with parameter θ:

$$
\begin{aligned}
\mathscr{L}(\theta|X) &= P_\theta(X_1 = x_1, X_2 = x_2, \ldots, X_N = x_N) \\
&= f(x_1|\theta) \cdot f(x_2|\theta) \ldots f(x_N|\theta) \\
&= \prod_{i=1}^{N} f(x_i|\theta).
\end{aligned}
\tag{2.36}
$$

The parameter θ can be estimated as the one that maximizes this likelihood:

$$
\theta^* = \arg\max_\theta \mathscr{L}(\mathscr{X}, \theta).
\tag{2.37}
$$

Given that the logarithm is a monotonic[p] function, maximizing the log of the likelihood produces the same as maximizing the likelihood itself. Therefore, we define a log-likelihood as:

$$
L(\theta|X) = \log(\mathscr{L}(\theta|X)) = \sum_{i=1}^{N} \log[f(x_i|\theta)].
\tag{2.38}
$$

The sensitivity of the log-likelihood for a specific parameter θ is known as the *score function*.

> **Example 2.2.2.** Normally Distributed Data
>
> A normally distributed random variable with mean μ and variance σ^2 has a distribution given by:
>
> $$
> f_X(x) = \frac{1}{\sqrt{2\pi\sigma^2}} \exp\left(-\frac{(x-\mu)^2}{2\sigma^2}\right).
> \tag{2.39}
> $$
>
> For a set of N i.i.d. observations $\mathbf{x} = \{x_1, \ldots, x_N\}$, the log-likelihood function is given by:
>
> $$
> L(\mu, \sigma^2|\mathbf{x}) = -\frac{N}{2}\log(2\pi\sigma^2) - \frac{1}{2\sigma^2}\sum_{i=1}^{N}(x_i - \mu)^2.
> \tag{2.40}
> $$
>
> Taking its derivative with respect to μ gives an estimation for μ:
>
> $$
> \frac{1}{\sigma^2}\sum_{i=1}^{N}(x_i - \mu) = 0 \rightarrow \hat{\mu} = \frac{1}{N}\sum_{i=1}^{N} x_i = \bar{x}.
> \tag{2.41}
> $$
>
> This states that the sample mean is the maximum likelihood estimator for the parameter μ.

---

[p]Given $a > b$, then $\log(a) > \log(b)$.

Similarly, calculating the derivative of the log-likelihood function for σ^2 gives:

$$-\frac{N}{2\sigma^2} + \frac{1}{2(\sigma^2)^2} \sum_{i=1}^{N}(x_i - \mu)^2 = 0 \rightarrow \hat{\sigma}^2 = \frac{1}{N}\sum_{i=1}^{N}(x_i - \bar{x})^2. \qquad (2.42)$$

■

2.3   KERNEL DENSITY ESTIMATION

Given a probability density, we can find the probability that a data point is in an interval $[a, b]$ by calculating:

$$P(a \le t \le b) = \int_a^b f(t)dt. \qquad (2.43)$$

The probability of finding a point inside a window $[x - h, x + h]$ for any arbitrary point x in an interval of interest can be written as:

$$P(x - h \le t \le x + h) = \int_{x-h}^{x+h} f(t)dt \approx 2hf(x). \qquad (2.44)$$

Consequently,

$$\begin{aligned} f(x) &\approx \frac{P(x - h \le t \le x + h)}{2h} \\ &\approx \frac{1}{2h}\frac{\#\text{ points } \in [x - h; x + h]}{N} \\ &\approx \frac{1}{Nh}\sum_{n=1}^{N} K_0\left(\frac{x - x_n}{h}\right), \end{aligned} \qquad (2.45)$$

where

$$K_0(u) = \begin{cases} 0, & \text{p/ } |u| \ge 1 \\ 1/2, & \text{p/ } |u| < 1 \end{cases} \qquad (2.46)$$

is a *kernel*. In Julia, this can be implemented as:

```
for xo in bins
    s = 0
    for xi in x
        u = (xi - xo)/h
        s += kernel(u)
    end
    push!(z,s/(N*h))
end
```

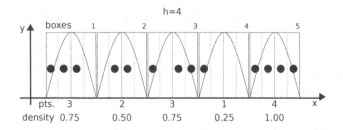

Figure 2.2 An example of computing a histogram. The circles correspond to experimental data, whereas the curves represent an Epanechnikov [1] window. The size of the boxes in all cases is $h = 4$

This computation is shown visually in Fig. 2.2. There are three points inside the first box. Since its width is 4, the density of points is $3/4$.

Calculating the histogram using this particular kernel, we observe that any point inside the window contributes equally to the density, whether the point is very close to the center of the box or close to one edge. To avoid this abrupt discontinuity it is possible to use different windows[q] [20, 21] that give different weights to points depending on where they are.

The bandwidth h is one of the most important parameters for estimating the density. A small bandwidth can produce a noise estimate, whereas a big bandwidth can create overly smoothed estimates as shown in Fig. 2.3.

The kernel used for estimating the density must have three essential properties: i) its zeroth moment must be unitary so that the results can be treated as probability density functions, ii) its first moment must be symmetric around 0 to preserve the average of the sample, and iii) its second moment must be positive:

$$\begin{cases} \int K(x)dx = 1 \\ \int xK(x)dx = 0 \\ \int x^2 K(x)dx > 0. \end{cases} \tag{2.47}$$

Some of the most common kernels, listed in Tab. 2.1, are given by the polynomial[r]:

$$K_s(u) = \frac{(2s+1)!!}{2^{s+1}s!}(1-u^2)^s \chi_{[-1;1]}(u), \tag{2.48}$$

2.3.1 BANDWIDTH

The simplest way to estimate the bandwidth is by computing the *mean integrated squared error* (MISE). Since $\hat{f}(x)$ and $f(x)$ are independent, $\langle \hat{f}(x)f(x) \rangle =$

---

[q]Emmanuel Parzen (1929–2016) American statistician.

[r]$\chi_I(x)$ is the indicator function that returns 1 if $x \in I$ and 0 otherwise.

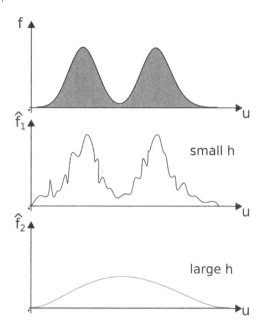

Figure 2.3   The statistical distribution of data (top), one estimate using a small bandwidth (center), and another estimate using a big bandwidth (bottom)

$\langle \hat{f}(x)\rangle \langle f(x)\rangle$. Therefore,

$$
\begin{aligned}
MISE(h) &= \left\langle \int (\hat{f}(x) - f(x))^2 dx \right\rangle \\
&= \int \langle \hat{f}^2(x) - 2\hat{f}(x)f(x) + f^2(x)\rangle \, dx \\
&= \int \left(\langle \hat{f}(x)\rangle - \langle f(x)\rangle \right)^2 + \langle \hat{f}^2(x)\rangle - \langle \hat{f}(x)\rangle^2 \, dx \\
&= \int BIAS^2\{\hat{f}(x)\} + VAR\{\hat{f}(x)\} dx
\end{aligned}
\tag{2.49}
$$

To estimate the bias, we can take the definition of a kernel:

$$
\begin{aligned}
\langle \hat{f}(x)\rangle &= \int \frac{1}{Nh} \sum_i^N K\left(\frac{x_i - t}{h}\right) f(t) dt \\
&= \frac{1}{N} \sum_i^N \int \frac{1}{h} K\left(\frac{x_i - t}{h}\right) f(t) dt.
\end{aligned}
\tag{2.50}
$$

Let's do the following substitution of variables:

$$
\begin{aligned}
u &= \frac{x_i - t}{h} \; \rightarrow \; t = x_i - hu, \\
du &= -\frac{1}{h} dt \; \rightarrow \; dt = -hdu.
\end{aligned}
\tag{2.51}
$$

Table 2.1

Values of $K_s(u)$ for $|u| \leq 1$.

$$
\begin{array}{rl}
\text{Uniform } (s = 0): & 0.5 \\
\text{Epanechnikov } (s = 1): & \tfrac{3}{4}(1 - u^2) \\
\text{Biweight } (s = 2): & \tfrac{15}{16}(1 - u^2)^2 \\
\text{Triweight } (s = 3): & \tfrac{35}{32}(1 - u^2)^3 \\
\text{Gaussian } (s \to \infty): & \tfrac{1}{\sqrt{2\pi}} \exp\left(-\tfrac{u^2}{2}\right)
\end{array}
$$

Therefore:

$$
\langle \hat{f}(x) \rangle = \frac{1}{N} \sum_i^N \int K(u) f(x_i - hu) du. \tag{2.52}
$$

The function $f(x_i - hu)$ can be Taylor expanded as:

$$
f(x_i - hu) = f(x_i) - hu f'(x_i) + \tfrac{1}{2} h^2 u^2 f''(x_i) + \ldots, \tag{2.53}
$$

Using it now back in Eq. 2.52, we get:

$$
\begin{aligned}
\langle \hat{f}(x) \rangle &\approx \frac{1}{N} \sum_i^N \int K(u) \left[f(x_i) - hu f'(x_i) + \tfrac{1}{2} h^2 u^2 f''(x_i) \right] du \\
&\approx \frac{1}{N} \sum_i^N \left[f(x_i) \int K(u) du - h f'(x_i) \int u K(u) du \right. \\
&\qquad\qquad \left. + \tfrac{1}{2} h^2 f''(x_i) \int u^2 K(u) du \right] \\
&\approx \frac{1}{N} \sum_i^N \left[f(x_i) + \tfrac{1}{2} h^2 f''(x_i) \sigma_K^2 \right]
\end{aligned}
\tag{2.54}
$$

$$
\langle \hat{f}(x) \rangle - \langle f(x) \rangle \approx \tfrac{1}{2} h^2 \sigma_K^2 \langle f''(x) \rangle = \tfrac{1}{2} h^2 \sigma_K^2 f''(x)
$$

$$
BIAS^2\{\hat{f}(x)\} \approx \tfrac{1}{4} h^4 \sigma_K^4 \left(f''(x) \right)^2.
$$

We see that the bias is proportional to the second moment of the kernel:

$$
\sigma_K^2 = \int_{-1}^{1} u^2 K(u) du = 2 \int_0^1 u^2 K(u) du. \tag{2.55}
$$

The second moments of some kernels are shown in 2.2.

The variance needed to estimate the MISE can be estimated as:

$$
\begin{aligned}
VAR\{\hat{f}(x)\} &= \langle \hat{f}^2(x) \rangle - \langle \hat{f}(x) \rangle^2 \\
&\approx \langle \hat{f}^2(x) \rangle - \left(\langle f(x) \rangle + \tfrac{1}{2} h^2 \sigma_K^2 f''(x) \right)^2
\end{aligned}
\tag{2.56}
$$

Table 2.2

Second Moment of Some Kernels (σ_K^2)

Kernel	σ_K^2
Uniformn: $2\int_0^1 u^2 \frac{1}{2} du = \frac{u^3}{3}\Big|_0^1$	$= \frac{1}{3}$
Epanechnikov: $2\int_0^1 u^2 \frac{3}{4}(1-u^2)du = \frac{3}{2}\left(\frac{u^3}{3} - \frac{u^5}{5}\Big|_0^1\right.$	$= \frac{1}{5}$
Quartic: $2\int_0^1 u^2 \frac{15}{16}(1-u^2)^2 du = 2\frac{15}{16}\left(\frac{u^3}{3} - 2\frac{u^5}{5} + \frac{u^7}{7}\Big|_0^1\right.$	$= \frac{1}{7}$
Triweight: $2\int_0^1 u^2 \frac{35}{32}(1-u^2)^3 du = 2\frac{35}{32}\left(\frac{u^6}{6} - 3\frac{u^5}{5} + 3\frac{u^4}{4} - u\Big|_0^1\right.$	$= \frac{1}{9}$
Gaussian: $\int_{-\infty}^{\infty} u^2 \frac{1}{\sqrt{2\pi}}\exp\left(-\frac{u^2}{2}\right)du$	$= 1$

The first term in the right-hand side of the equation is:

$$
\langle \hat{f}^2(x) \rangle = \int \frac{1}{N^2 h^2}\left(\sum_i K\left(\frac{x_i - t}{h}\right)\right)^2 f(t)dt
$$

$$
= \frac{1}{N^2 h^2}\int \left[\sum_i K\left(\frac{x_i - t}{h}\right)\right]\left[\sum_j K\left(\frac{x_j - t}{h}\right)\right]f(t)dt
$$

(2.57)

Kernels are centered in different points x_α. Therefore, their products are non-zero only when $i = j$. Consequently:

$$
\langle \hat{f}^2(x) \rangle = \frac{1}{N^2 h^2}\sum_i \int K^2\left(\frac{x_i - t}{h}\right)f(t)dt
$$

(2.58)

Using the same substitution of variables from Eq. 2.51, we get:

$$
\langle \hat{f}^2(x) \rangle = \frac{1}{N^2 h}\sum_i \int K^2(u)f(x_i - hu)du
$$

$$
\approx \frac{1}{N^2 h}\sum_i \int K^2(u)\left[f(x_i) - huf'(x_i)\right]du
$$

$$
\approx \frac{1}{Nh}\frac{1}{N}\sum_i f(x_i)\int K^2(u)du
$$

(2.59)

$$
\approx \frac{1}{Nh}\langle f(x) \rangle R_k,
$$

where $R_k = \int K^2(x)dx$ is the roughness of the kernel.

The variance then becomes:

$$VAR\{\hat{f}(x)\} = \langle \hat{f}^2(x) \rangle - \langle \hat{f}(x) \rangle^2$$

$$\approx \frac{1}{Nh} \langle f(x) \rangle R_k - \left(\langle f(x) \rangle + \frac{1}{2} h^2 \sigma_K^2 f''(x) \right)^2 \qquad (2.60)$$

$$\approx \frac{1}{Nh} \langle f(x) \rangle R_k - \langle f(x) \rangle^2.$$

When the bandwidth is small, the first term tends to be much higher than the second. Therefore, the variance can be approximated as:

$$VAR\{\hat{f}(x)\} \approx \frac{1}{Nh} \langle f(x) \rangle R_k. \qquad (2.61)$$

Finally, an asymptotic version of the MISE (AMISE) can be written as:

$$AMISE = \int BIAS^2\{\hat{f}(x)\} + VAR\{\hat{f}(x)\} dx$$

$$\approx \int \left[\frac{1}{4} h^4 \sigma_K^4 (f''(x))^2 + \frac{1}{Nh} \langle f(x) \rangle R_k \right] dx \qquad (2.62)$$

$$\approx \frac{1}{4} h^4 \sigma_K^4 R_{f''} + \frac{R_k}{Nh}.$$

The bandwidth that minimizes the AMISE is given by:

$$\frac{dAMISE}{dh} = h^3 \sigma_k^4 R_{f''} - \frac{R_k}{Nh^2} = 0$$

$$\therefore h = \left(\frac{R_k}{N \sigma_k^4 R_{f''}} \right)^{1/5}. \qquad (2.63)$$

The bandwidth depends on the unknown $R_{f''}$, which quantifies how much the distribution oscillates. Therefore, the more high-frequency components the distribution has, the smaller the bandwidth must be to avoid averaging the oscillations instead of following them.

To make calculations simpler, we can take an arbitrary distribution and do the following trick:

$$f_\sigma(u) = \frac{1}{\sigma} f \left(\frac{u}{\sigma} \right)$$

$$f_\sigma''(u) = \sigma^{-3} f'' \left(\frac{u}{\sigma} \right). \qquad (2.64)$$

The roughness can now be estimated as:

$$(R_{f_\sigma''})^{-1/5} = \left(\int (f_\sigma''(u))^2 du \right)^{-1/5}$$

$$= \left(\int \sigma^{-6} \left(f'' \left(\frac{u}{\sigma} \right) \right)^2 du \right)^{-1/5}. \qquad (2.65)$$

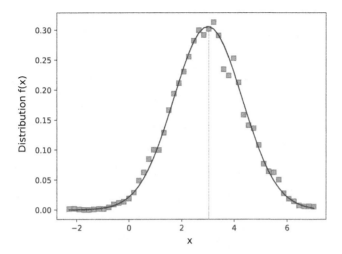

Figure 2.4 Theoretical normal distribution $\mathcal{N}(3.0, 1.7)$ (solid curve) and the estimated distribution for 50 bins using a quartic kernel (squares)

Substituting $x = u/\sigma$ and $dx = du/\sigma$, we obtain:

$$\left(R_{f''_\sigma}\right)^{-1/5} = \left(\int \sigma^{-5}\left(f''(x)\right)^2 dx\right)^{-1/5}$$

$$= \sigma\left(R_{f''}\right)^{-1/5}. \tag{2.66}$$

Therefore, the bandwidth can be rewritten using the results from Eqs. 2.63 and 2.66:

$$h = C\sigma N^{-1.5}, \tag{2.67}$$

where

$$C = \left(\frac{R_k}{N\sigma_k^4 R_{f''}}\right)^{1/5}. \tag{2.68}$$

For the normal distribution, it can be shown that the bandwidth can be approximated by:

$$h = 1.06\sigma N^{-1/5}. \tag{2.69}$$

Figure 2.4 shows the real normal distribution centered in $\mu = 3.0$ with a variance $\sigma^2 = 1.7$ and the estimated distribution using a quartic (or biweight) kernel. Data was composed of $N = 2000$ points and the kernel was computed using 50 bins.

2.4 MARKOV CHAINS

A Markov[s] chain [22, 23] is a stochastic process in which the probability of transitioning to a new state depends solely on the current state, a memoryless principle known as the *Markov property* [24].

---

[s]Andrey Andreyevich Markov (1856–1922) Russian mathematician, advised by Pafnuty Chebyshev, and adviser of Abram Besicovitch and Georgy Voronoy, among others.

To explore this concept in more depth, let's take a closer look at its mathematical underpinnings. Consider a stochastic process $X = \{X_t : t \in \mathbb{N}\}$ defined on a probability space $(\Omega, \mathscr{F}, P)$ with a discrete state space $(S, \mathscr{S}), S \subseteq \Omega$, where $\mathscr{S}$ is a σ-algebra of the state space S. This process is a Markov chain if it satisfies the condition $P(X_{t+1} = x | X_0 = x_0, X_1 = x_1, \ldots, X_t = y) = P(X_{t+1} = x | X_t = y)$, where $x, x_0, x_i, y \in S$.

The probability $P : \Omega \times \mathscr{F} \to \mathbb{R}$ is the probability transition function, which is organized in a transition matrix $P = (p_{ij})_{S \times S}$ provided that S is countable. For example, consider two states: *hot* and *cold* represented by the transition probability graph in Fig. 2.5.

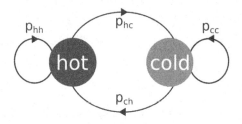

Figure 2.5   A Markov chain with two states: *hot* and *cold*

The transition matrix corresponding to this Markov chain is given by:

$$\mathbf{P} = \begin{bmatrix} p_{hh} & p_{hc} \\ p_{ch} & p_{cc} \end{bmatrix} \tag{2.70}$$

Many times, we want to know the probability of finding the system at some particular state in the future. For this, consider that the initial state is i and we ask what is the probability of finding the system at state j after n steps:

$$P(X_n = j | X_0 = i) = \sum_{k=1}^{m} P(X_{n-1} = k | X_0 = i) P(X_n = j | X_{n-1} = k, X_0 = i)$$

$$r_{ij}(n) = \sum_{k=1}^{m} r_{ik}(n-1) p_{kj}, \tag{2.71}$$

where $r_{ij}(n)$ is the probability that given an initial state i, state j will be reached after n steps. This recursion relationship is known as the Chapman[t]-Kolmogorov[u] equation [9].

If the Markov chain is aperiodic and irreducible, implying that every state can be reached from every other state with some positive probability, then the Markov chain will converge to a steady-state behavior where the probabilities related to different states do not change further over time. The probability distribution π represents this

[t]Sydney Chapman (1888–1970) British mathematician advised by Godfrey Harold Hardy. Chapman was the winner of many prizes, including the Copley medal in 1964.
[u]Andrey Nikolaevich Kolmogorov (1903–1987) Russian mathematician.

long-term behavior and is called the *stationary distribution*, which can be found by solving the Chapman-Kolmogorov equation:

$$\lim_{n\to\infty} r_{ij}(n) = \sum_{k=1}^{m} \lim_{n\to\infty} r_{ik}(n-1)p_{kj}$$

$$\pi_{ij} = \sum_{k=1}^{m} \pi_{ik}p_{kj} \tag{2.72}$$

$$\pi = \pi\mathbf{P}.$$

Let's now consider our initial Markov chain $\ldots,X_{n-2},X_{n-1},X_n,\ldots$ and let's define a stochastic process that runs backward $\ldots,X_n,X_{n-1},X_{n-2},\ldots$. This is also a Markov chain because the Markov property states that the current state depends on the previous state regardless if it is moving forward or backward in time. The transition probability, in this case, is given by:

$$\begin{aligned}
Q_{ij} &= P(X_m = j | X_{m+1} = i) \\
&= \frac{P(X_m = j, X_{m+1} = i)}{P(X_{m+1} = i)} \\
&= \frac{P(X_m = j)P(X_{m+1} = i | X_m = j)}{P(X_{m+1} = i)} \\
&= \frac{\pi_j P_{ji}}{\pi_i}.
\end{aligned} \tag{2.73}$$

If the Markov chain X is reversible, then Q_{ij} must be the same as P_{ij} and consequently, we obtain:

$$\pi_j P_{ji} = P_{ij}\pi_i. \tag{2.74}$$

This condition is known as the *detailed balance* equation. If π is a solution to this equation, then it is unique, and to show this we compute:

$$\sum_i \pi_i P_{ij} = \sum_i \pi_j P_{ji} = \pi_j \sum_i P_{ji} = \pi_j, \tag{2.75}$$

which is the stationary distribution.

2.4.1 HIDDEN MARKOV MODELS

Imagine now that you need to study the climate of some region years ago but there is no available information about the temperature of that region. However, you found some records about the number of tickets for a ski resort in that region. Given a sequence O of observations of the number of tickets sold $\{Few, Many\}$, how can we infer a hidden sequence X of weather states $\{Hot, Cold\}$? This new problem where observations are assumed to be generated by an underlying hidden process is illustrated in the transition probability graph of Fig. 2.6. It has important applications where sequences of data need to be analyzed including speech recognition, robot localization, and gene identification.

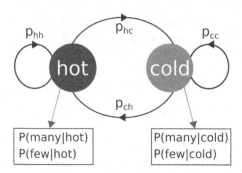

Figure 2.6 A Markov chain with two states *hot* and *cold* and the respective obser-vations for each state

The pointed-out problem is one of three that can be solved with hidden Markov models (HMM) [25, 26]:

i) Evaluation: Given an HMM with known parameters, what is the probability of observing a specific sequence?

ii) Decoding: Given an HMM with known parameters and a sequence of obser-vations, what is the most probable sequence of hidden states?

iii) Learning: Given an HMM with unknown parameters, and a sequence of ob-servations, what are the parameters that maximize the likelihood of observed data?

2.4.1.1 Evaluation

For a sequence of hidden states $X = \{X_1, \ldots, X_T\}$ and observations $O = \{O_1, \ldots, O_T\}$, the likelihood of obtaining this sequence of observations is:

$$P(O|X) = \prod_{i=1}^{T} P(O_i|X_i), \tag{2.76}$$

where the probability of generating an observation O_i given the current state X_i is known as *emission probability*. The problem is, however, that we do not know the hidden states. We can get around this problem by weight-summing over all possible states:

$$P(O) = \sum_X P(O, X) = \sum_X P(O|X)P(X)$$

$$= \sum_{X_1, \ldots, X_T} P(X_1) \left(\prod_{i=1}^{T} P(O_i|X_i) \right) \left(\prod_{i=2}^{T} P(X_i|X_{i-1}) \right). \tag{2.77}$$

Note that here we applied the chain rule and the Markov property: $P(X) = \prod_{i=1}^{T} P(X_i|X_1, \ldots, X_{i-1}) = \prod_{i=1}^{T} P(X_i|X_{i-1})$.

This is computationally costly. Instead, the probability of observing a sequence can be written as:

$$P\left(\{O_t\}_{t=1}^T\right) = \sum_k P\left(\{O_t\}_{t=1}^T, X_T = k\right) = \sum_k \alpha_T^k, \qquad (2.78)$$

where

$$\begin{aligned}
\alpha_t^k &= P\left(\{O_r\}_{r=1}^t, X_t = k\right) \\
&= P\left(\{O_r\}_{r=1}^t | X_t = k\right) P(X_t = k) \\
&= P(O_t | X_t = k) P\left(\{O_r\}_{r=1}^{t-1} | X_t = k\right) P(X_t = k) \\
&= P(O_t | X_t = k) P\left(\{O_r\}_{r=1}^{t-1}, X_t = k\right) \\
&= P(O_t | X_t = k) \sum_j P\left(\{O_r\}_{r=1}^{t-1}, X_t = k | X_{t-1} = j\right) P(X_{t-1} = j) \\
&= P(O_t | X_t = k) \sum_j P\left(\{O_r\}_{r=1}^{t-1} | X_{t-1} = j\right) P(X_t = k | X_{t-1} = j) P(X_{t-1} = j) \\
&= P(O_t | X_t = k) \sum_j P\left(\{O_r\}_{r=1}^{t-1}, X_{t-1} = j\right) P(X_t = k | X_{t-1} = j) \\
&= P(O_t | X_t = k) \sum_j \alpha_{t-1}^j P(X_t = k | X_{t-1} = j).
\end{aligned}$$

$$(2.79)$$

This recursive relation (see Sec. 10.1) is the basis for the *forward algorithm* [27]:

Algorithm 1: Forward Algorithm

$\forall k, \ \alpha_1^k = p(O_1 | X_1 = k) p(X_1 = k)$;
for $t = 2, \ldots, T$ **do**
$\quad \triangleright \forall k, \ \alpha_t^k = P(O_t | X_t = k) \sum_j \alpha_{t-1}^j P(X_t = k | X_{t-1} = j)$;
end
$\triangleright P\left(\{O_t\}_{t=1}^T\right) = \sum_k \alpha_T^k$;

2.4.1.2 Decoding

For the inverse problem of finding the probability for a specific hidden state $X_t = k$ given observations $O = \{O_1, \ldots, O_T\}$, we can set:

$$P\left(X_t = k | \{O_t\}_{t=1}^T\right) = \frac{P\left(\{O_t\}_{t=1}^T, X_t = k\right)}{P(\{O_t\}_{t=1}^T)} \qquad (2.80)$$

The denominator was just found using the forward algorithm. The numerator can be written as:

$$\begin{aligned}
P\left(\{O_r\}_{r=1}^T, X_t = k\right) &= P\left(\{O_r\}_{r=1}^t, X_t = k, \{O_r\}_{r=t+1}^T\right) \\
&= P(\{O_r\}_{r=1}^t, X_t = k) P(\{O_r\}_{r=t+1}^T | \{O_r\}_{r=1}^t, X_t = k) \\
&= P(\{O_r\}_{r=1}^t, X_t = k) P(\{O_r\}_{r=t+1}^T | X_t = k) \\
&= \alpha_t^k \beta_t^k.
\end{aligned}$$

$$(2.81)$$

The component β_t^k can be found recursively with:

$$
\begin{aligned}
\beta_t^k &= P(\{O_r\}_{r=t+1}^T | X_t = k) \\
&= \sum_j P(O_{t+1}, \{O_r\}_{r=t+2}^T, X_{t+1} = j, X_t = k)/P(X_t = k) \\
&= \sum_j P(O_{r=t+2}^T | X_{t+1} = j, X_t = k) P(O_{t+1}, X_{t+1} = j, X_t = k)/P(X_t = k) \\
&= \sum_j P(O_{r=t+2}^T | X_{t+1} = j) P(O_{t+1} | X_{t+1} = j, X_t = k) \frac{P(X_{t+1} = j, X_t = k)}{P(X_t = k)} \\
&= \sum_j \beta_{t+1}^j P(O_{t+1} | X_{t+1} = j) P(X_{t+1} = j | X_t = k).
\end{aligned}
\tag{2.82}
$$

This leads to the *backward algorithm* [27]:

Algorithm 2: Backward Algorithm

$\forall k, \beta_T^k = 1$;
for $t = T - 1, \ldots, 1$ **do**
$\quad \triangleright \forall k, \beta_t^k = \sum_j \beta_{t+1}^j P(O_{t+1} | X_{t+1} = j) P(X_{t+1} = j | X_t = k)$;
end
$\triangleright P\left(\{O_r\}_{r=1}^T, X_t = k\right) = \alpha_t^k \beta_t^k \to P\left(X_t = k | \{O_t\}_{t=1}^T\right) = \frac{\alpha_t^k \beta_t^k}{\sum_r \alpha_t^r \beta_t^r}$;

Note that we have used Eq. 2.81 to arrive at the last equation of the algorithm. Also, the backward algorithm finds the most likely state, not a state sequence. The latter can be found by solving the following equation:

$$
\arg\max_{\{X_t\}_{t=1}^T} P\left(\{X_t\}_{t=1}^T | \{O_t\}_{t=1}^T\right) = \arg\max_{\{X_t\}_{t=1}^T} \frac{P\left(\{X_t\}_{t=1}^T, \{O_t\}_{t=1}^T\right)}{P(\{O_t\}_{t=1}^T)}
\tag{2.83}
$$

Since we are computing for $\{X_t\}_{t=1}^T$, the denominator does not play any role and can be eliminated:

$$
= \arg\max_{\{X_t\}_{t=1}^T} P\left(\{X_t\}_{t=1}^T, \{O_t\}_{t=1}^T\right).
\tag{2.84}
$$

Now, the sequence $\{X_t\}_{t=1}^T$ that maximizes the probability can be decomposed to a sequence that ends with $X_T = k$:

$$
= \arg\max_k \left[\arg\max_{\{X_t\}_{t=1}^{T-1}} P\left(X_T = k, \{X_t\}_{t=1}^{T-1}, \{O_t\}_{t=1}^T\right) \right].
\tag{2.85}
$$

Just like we did in the previous algorithms, this equation can be computed recursively by setting:

$$
\begin{aligned}
V_t^k &= \arg\max_{\{X_r\}_{r=1}^{t-1}} P\left(X_t = k, \{X_r\}_{r=1}^{t-1}, \{O_r\}_{r=1}^t\right) \\
&= \arg\max_{\{X_t\}_{r=1}^{t-1}} P\left(O_t | \{O_r\}_{r=1}^{t-1}, X_t = k, \{X_r\}_{r=1}^{t-1}\right) P\left(\{O_r\}_{r=1}^{t-1}, X_t = k, \{X_r\}_{r=1}^{t-1}\right) \\
&= P(O_t | X_t = k) \arg\max_{\{X_t\}_{r=1}^{t-1}} P\left(X_t = k | \{X_r\}_{r=1}^{t-1}, \{O_r\}_{r=1}^{t-1}\right) P\left(\{X_r\}_{r=1}^{t-1}, \{O_r\}_{r=1}^{t-1}\right) \\
&= P(O_t | X_t = k) \arg\max_i P(X_t = k | X_{t-1} = i) V_{t-1}^i.
\end{aligned}
$$
(2.86)

With this, we arrive at the Viterbi[v] algorithm [28, 29]:

Algorithm 3: Viterbi Algorithm

$\forall k, \ V_1^k = P(O_1 | X_1 = k) P(X_1 = k);$
for $t = t = 2, \ldots, T$ **do**
$\quad \triangleright V_t^k = P(O_t | X_t = k) \arg\max_i P(X_t = k | X_{t-1} = i) V_{t-1}^i;$
end
$\triangleright \arg\max_{\{X_t\}_{t=1}^T} P\left(\{X_t\}_{t=1}^T, \{O_t\}_{t=1}^T\right) = \arg\max_k V_T^k;$

The states can be estimated backward with the traceback procedure:

$$
\begin{aligned}
X_T &= \arg\max_k V_T^k \\
X_{t-1} &= \arg\max_i P(X_t | X_{t-1} = i) V_{t-1}^i.
\end{aligned}
$$
(2.87)

2.4.1.3 Learning

The learning problem is about estimating the parameters of the HMM given an observation sequence $\mathbf{O} = \{O_t\}_{t=1}^T$. For this, let's parameterize the HMM with the tuple $\theta = (\pi, A, B)$, where the elements $\pi_i = P(X_1 = i)$ constitute the initial state vector, $A_{ij} = P(X_{t+1} = j | X_t = i)$ are the elements of the transition matrix, and $B_i(j) = P(O_t = j | X_t = i)$ are the elements of the emission matrix. In all cases, $\mathbf{X} = \{X_t\}_{t=1}^T$ is a sequence of hidden states. The problem now is to find:

$$
\theta^* = \arg\max_\theta \sum_{\mathbf{X}} P(\mathbf{O}, \mathbf{X} | \theta).
$$
(2.88)

---
[v]Andrea James Viterbi (1935–Pres.) Italian/American electrical engineer winner of many prizes, including the Claude E. Shannon Award in 1991.

Jensen's Inequality

A function $f : \mathbb{R} \to \mathbb{R}$ is convex if the tangent to any of its points x_0 lies below the graph of the function. Mathematically:

$$f(x) \geq f(x_0) + g(x_0)(x - x_0), \tag{2.89}$$

where $g(x_0) = \left.\frac{df(x)}{dx}\right|_{x=x_0}$ is the tangent of the curve at x_0.

Substituting x for an integrable[a] random variable X and x_0 for its expected value $\langle X \rangle$:

$$f(X) \geq f(\langle X \rangle) + g(\langle X \rangle)(X - \langle X \rangle). \tag{2.90}$$

Taking the expectation of the function:

$$\langle f(X) \rangle \geq \langle f(\langle X \rangle) \rangle + g(\langle X \rangle)\langle (X - \langle X \rangle) \rangle$$
$$\geq f(\langle X \rangle). \tag{2.91}$$

This is known as Jensen's[b] inequality [30].

---

[a] a random variable whose absolute summability is finite: $\sum_x |x| p_X(x) < \infty$, where $p_X(x)$ is its probability mass function.

[b] Johan Ludwig William Valdemar Jensen (1859–1925) Danish mathematician.

The log-likelihood for this joint probability distribution can be written as:

$$\log[P(\mathbf{O}|\theta)] = \log\left[\sum_{\mathbf{X}} P(\mathbf{O}, \mathbf{X}|\theta)\right]$$
$$= \log\left[\sum_{\mathbf{X}} P(\mathbf{X}|\theta') \frac{P(\mathbf{O}, \mathbf{X}|\theta)}{P(\mathbf{X}|\theta')}\right]. \tag{2.92}$$

Using Jensen's inequality:

$$\log[P(\mathbf{O}|\theta)] \geq \sum_{\mathbf{X}} P(\mathbf{X}|\theta') \log\left[\frac{P(\mathbf{O}, \mathbf{X}|\theta)}{P(\mathbf{X}|\theta')}\right]$$
$$\geq \sum_{\mathbf{X}} P(\mathbf{X}|\theta') \log[P(\mathbf{O}, \mathbf{X}|\theta)] - \sum_{\mathbf{X}} P(\mathbf{X}|\theta') \log[P(\mathbf{X}|\theta')] \tag{2.93}$$
$$\geq Q(\theta|\theta') + H(\theta').$$

The right-hand side of the equation is a *lower bound* on the likelihood. Function H is the entropy, whereas the Q-function is the expected complete log-likelihood.

Since the entropy does not depend on θ, it suffices to maximize the Q-function to maximize the log-likelihood. One way to accomplish this is using the

Baum[w]-Welch[x] algorithm [26, 27]. This is an algorithm that works by successively iterating between the computation of the expected statistics[y] of the HMM (E-step) and then maximizing the expected log-likelihood of the observations for the parameters θ (M-step). Mathematically, these steps can be written as:

$$Q(\theta|\theta^n) = \sum_{\mathbf{X}} P(\mathbf{X}|\mathbf{O},\theta^n)\log[P(\mathbf{O},\mathbf{X}|\theta)].$$

$$\theta^{n+1} = \arg\max_{\theta} Q(\theta|\theta^n) \tag{2.94}$$

E-step

To compute the log-likelihood, we can use the chain rule:

$$Q(\theta|\theta^n) = \sum_{\mathbf{X}} P(\mathbf{X}|\mathbf{O},\theta^n)\log[P(\mathbf{O},\mathbf{X}|\theta)]$$

$$= \sum_{\mathbf{X}} P(\mathbf{X}|\mathbf{O},\theta^n)\log\left[P(X_1|\theta)P(O_1|X_1,\theta)\prod_{t=2}^{T}P(O_t|X_t,\theta)P(X_t|X_{t-1},\theta)\right]$$

$$= \sum_{\mathbf{X}} P(\mathbf{X}|\mathbf{O},\theta^n)\log[P(X_1|\theta)] +$$

$$+ \sum_{\mathbf{X}}\sum_{t=2}^{T} P(\mathbf{X}|\mathbf{O},\theta^n)\log[P(X_t|X_{t-1},\theta)] +$$

$$+ \sum_{\mathbf{X}}\sum_{t=1}^{T} P(\mathbf{X}|\mathbf{O},\theta^n)\log[P(O_t|X_t,\theta)].$$

$$\tag{2.95}$$

This equation can be rewritten using only the components of the model that each term depends on:

$$Q(\theta|\theta^n) = \langle\log[P(X_1|\theta)]\rangle_{X_1}$$

$$+ \sum_{t=2}^{T}\langle\log[P(X_t|X_{t-1},\theta)]\rangle_{X_t,X_{t-1}} \tag{2.96}$$

$$+ \sum_{t=1}^{T}\langle\log[P(O_t|X_t,\theta)]\rangle_{X_t}.$$

We can take a step further using our time-invariant definitions:

$$Q(\theta|\theta^n) = \langle\pi_0\rangle_{X_1} + \langle\log(A_{ij})\rangle_{X_t,X_{t-1}} + \langle\log(B_i(j))\rangle_{X_t}. \tag{2.97}$$

This decomposition implies that each matrix can be estimated individually. The sufficient statistics, or *responsibilities*, (see Appendix A) for this problem are expected

---

[w]Leonard Esau Baum (1931–2017) American mathematician.

[x]Lloyd Richard Welch (1927–Pres.) American mathematician, winner of the Claude E. Channon Award in 2003.

[y]Statistics that contain all information necessary to estimate the parameters of the model: the expected number of transitions, the expected number of times a state is visited, and the expected number of emissions.

counts of the hidden states γ_t^k and the expected number of transitions between them, given the observed sequence ξ_t^k.

The expected number of times in each state is given by:

$$
\begin{aligned}
\gamma_t^k = P(X_t = k|\mathbf{O}) &= \frac{P(X_t = k, \mathbf{O})}{P(\mathbf{O})} \\
&= \frac{P(\mathbf{O}|X_t = k)P(X_t = k)}{P(\mathbf{O})} \\
&= \frac{P(\{O_r\}_{r=1}^t|X_t = k)P(\{O_r\}_{r=t+1}^T|X_t = k)P(X_t = k)}{P(\mathbf{O})} \\
&= \frac{P(\{O_r\}_{r=1}^t, X_t = k)P(\{O_r\}_{r=t+1}^T|X_t = k)}{P(\mathbf{O})} \\
&= \frac{\alpha_t^k \beta_t^k}{\sum_j \alpha_t^j \beta_t^j}.
\end{aligned}
\tag{2.98}
$$

The expected number of transitions between states, given an observed sequence, is, successively applying conditionality, given by:

$$
\begin{aligned}
\xi_t^{ij} = P(X_t = i, X_{t+1} = j|\mathbf{O}) &\\
&= \frac{P(X_t = i, X_{t+1} = j, \{O_r\}_{r=1}^t, O_{t+1}, \{O_r\}_{r=t+2}^T)}{P(\mathbf{O})} \\
&= \frac{\beta_{t+1}^i B_j(O_{t+1})\alpha_t^i A_{ij}}{P(\mathbf{O})}.
\end{aligned}
\tag{2.99}
$$

M Step

After taking the derivatives of the Q-function and setting them to zero, updating the parameters of the HMM come directly from sufficient statistics. The initial state vector is updated as the expected frequency spent in state i at the first instant:

$$
\pi_i^{k+1} = \gamma_1^i.
\tag{2.100}
$$

The transition matrix is updated by the expected number of transitions between states normalized by the total number of transitions departing from the initial state:

$$
A_{ij}^{k+1} = \frac{\sum_{t=1}^{T-1} \xi_t^{ij}}{\sum_{t=1}^{T-1} \gamma_t^i}.
\tag{2.101}
$$

Finally, the emission matrix is updated as the expected number of times the observations match the expectation:

$$
B_i^{k+1}(W_k) = \frac{\sum_{t=1}^T \gamma_t^i 1_{O_t = W_k}}{\sum_{t=1}^T \gamma_t^i},
\tag{2.102}
$$

where $1_{O_t = W_k}$ is the indicator function that returns 1 only when the condition $O_t = W_k$ is satisfied.

Section II

Unsupervised Learning

3 Dimensionality Reduction

This is the first chapter about unsupervised learning, a collection of methods to discover patterns, structures, and relationships in unlabeled data. This first chapter is about dimensionality reduction, which aims at reducing the number of features in a dataset while preserving its essential information. This has some useful applications such as removing noise from the data.

First, we will discuss principal component analysis, a linear dimensionality reduction technique, that is widely used for data compression. Next, we will discuss singular value decomposition, which is a more general concept and provides a general framework for matrix factorization. We will then discuss a kernel trick to work with nonlinear features in data. We will end the chapter with a technique that tries to separate a multi-component signal into its underlying independent components.

3.1 PRINCIPAL COMPONENT ANALYSIS

Imagine that you have an industrial process that is monitored by three sensors. Let's label them sensors x, y, and z and create a vector $\mathbf{r} = \begin{bmatrix} x & y & z \end{bmatrix}^\top \in \mathbb{R}^3$ that contains these outputs. These sensors produce outputs every minute so that after one hour we will have 60 vectors $\mathbf{r}_i = \begin{bmatrix} x_i & y_i & z_i \end{bmatrix}^\top$, $i = 1, \ldots, 60$. The dimension of our data set is $d = 3$ and the number of points is $N = 60$. To organize our data, we will create a matrix $\mathbf{R} = \begin{bmatrix} \mathbf{r}_1 & \mathbf{r}_2 & \ldots & \mathbf{r}_N \end{bmatrix} \in \mathbb{R}^{d \times N}$.

Now, we can ask how these data points are organized or whether they cluster together or are dispersed. Principal component analysis (PCA) is a technique developed by Pearson[a] [31, 32] that tries to offer some answers to these questions.

Before beginning this study, we must be aware that the data sets might be dispersed in different ranges. Therefore, to be able to study them properly, we must rescale them to the same range by doing:

$$\text{row}_i \mathbf{R} \mapsto \frac{\text{row}_i \mathbf{R} - \langle \text{row}_i \mathbf{R} \rangle}{\sigma \left(\text{row}_i \mathbf{R} \right)}, \tag{3.1}$$

where σ indicates the standard variation of the dataset. In Julia, the normalization for a matrix R is written as:

```
using Statistics

R = (R .- mean(R,dims=2)) ./ std(R,dims=2)
```

[a]Karl Pearson (1857–1936) English mathematician, advised by Francis Galton, and adviser of John Wishart, among others. The contributions of Pearson to statistics are numerous. Correlation analysis and histograms, for instance, are credited to Pearson. He was awarded many prizes including a Fellowship by the Royal Society in 1896.

DOI: 10.1201/9781003350101-3

3.1.1 CORRELATION MATRIX

The data are often correlated, but we would like to find the axes onto which it is most dispersed. Therefore, we assume that this new coordinate system $\mathbf{Y} = \begin{bmatrix} \mathbf{y}_1 & \mathbf{y}_2 & \cdots & \mathbf{y}_N \end{bmatrix} \in \mathbb{R}^{d \times N}$ can be written as a linear combination on the data set:

$$\mathbf{Y} = \mathbf{VR}, \tag{3.2}$$

where $\mathbf{V} \in \mathbb{R}^{d \times d}$ is a matrix of weights.

The dispersion along those axes is given by the correlation matrix:

$$\begin{aligned}
\text{corr}(\mathbf{Y}) &= \frac{1}{N-1}\mathbf{YY}^{\top} \\
&= \mathbf{V}\left(\frac{1}{N-1}\mathbf{RR}^{\top}\right)\mathbf{V}^{\top} \\
&= \mathbf{VCV}^{\top},
\end{aligned} \tag{3.3}$$

where $\mathbf{C} \in \mathbb{R}^{d \times d}$ is the correlation matrix of the normalized dataset. Note that this can also be written as:

$$\text{corr}(\mathbf{Y})_{ij} = (\text{row}_i\mathbf{V})\,\mathbf{C}\left(\text{col}_j\mathbf{V}^{\top}\right) = \mathbf{v}_i^{\top}\mathbf{C}\mathbf{v}_j. \tag{3.4}$$

Lagrange Multipliers

Imagine that you are in the desert and want to move from point A to point B. However, you need to stop by a river to get some water to help your journey. This is depicted in the figure below as a solid line. Since water is scarce, you want to make your journey the shortest possible.

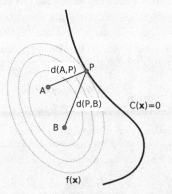

The loci of equidistant points from A and B are given by ellipses indicated by dotted lines whereas the curve that represents the river is given by $C(\mathbf{x}) = C_0$. Thus, the cost function $f(\mathbf{x})$ is just the path length $d(A,P) + d(P,B)$, where P is a point along the stream. Graphically, we see that the solution passes through a point P that is tangent to the equidistance curves. Mathematically,

> ## Lagrange Multipliers (Continued)
>
> this is achieved when both normal vectors are equal in magnitude but in opposite directions:
>
> $$\nabla f(\mathbf{x}) = \lambda \nabla C(\mathbf{x})$$
> $$\nabla \left(f(\mathbf{x}) - \lambda C(\mathbf{x}) \right) = 0, \tag{3.5}$$
>
> where λ is a proportionality constant called the *Lagrange[a] multiplier* [33]. Therefore, we can understand the above equation as a minimization of the *Lagrange function*:
>
> $$\Lambda = f(\mathbf{x}) - \lambda \left(C(\mathbf{x}) - C_0 \right). \tag{3.6}$$
>
> ---
>
> [a]Joseph-Louis Lagrange (1736–1813) Italian mathematician, advised by Leonhard Euler, and adviser of Joseph Fourier and Siméon Poisson, among others.

We would like the vectors $\mathbf{v}_n$ to be unitary so that they only indicate directions. Furthermore, we would like this new coordinate system to have uncorrelated axes so that we would be able to know what independent directions give the most variability of the data. Therefore, the correlation matrix[b] should only have nonzero elements in the main diagonal. Now, to maximize the variance and find the axes of most data variability under these conditions, we can maximize a Lagrangian function:

$$\frac{\partial}{\partial \mathbf{v}_n} \left(\mathbf{v}_n^\top \mathbf{C} \mathbf{v}_n - \lambda_n \left[\mathbf{v}_n^\top \mathbf{v}_n - 1 \right] \right) = 0 \tag{3.7}$$
$$\mathbf{C} \mathbf{v}_n = \lambda_n \mathbf{v}_n.$$

This is an eigenstate equation whose eigenvalues λ_n give the scaling factors of the eigenvectors (or principal components) $\mathbf{v}_n$. While the former also corresponds to the variance explained, the latter indicates the directions along which most of the data variability occurs. Furthermore, the set of eigenvectors forms an orthogonal basis for the analyzed data[c]. Therefore, the new basis is uncorrelated. Given this description, the highest eigenvalue corresponds to a principal component along which most of the data covariability occurs. Each of the subsequent eigenvalues indicates a direction of diminishing data covariability. In our example, this is shown in Fig. 3.1.

In Julia, we can create a correlation matrix and find its eigenstates with:

```
using LinearAlgebra

C = R*R'/(N-1)

va = eigvals(C)
ve = eigvecs(C)
```

---

[b]Note that the correlation matrix is symmetric.
[c]This technique is often called the *orthogonal linear transformation*.

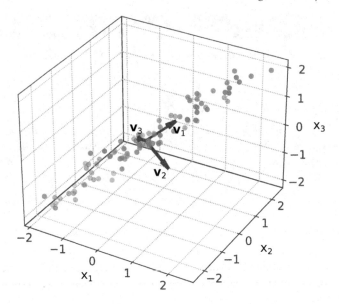

Figure 3.1 Three-dimensional display of experimental data (disks). The black arrows are the eigenvectors of the correlation matrix and indicate the axes of maximum variability of data

3.1.2 EXPLAINED VARIANCE

Since the correlation matrix is positive definite, all its eigenvalues are positive. Therefore, with the eigenvalues sorted in descending order, the *proportion of explained variance* (PEV) is the ratio between the variance of each component and the total variance:

$$\sigma_{n,\%}^2 = \frac{\lambda_n}{\sum_m \lambda_m}. \tag{3.8}$$

A plot of PEV as a function of the eigenvalue number is typically known as a *scree plot*. This usually has the shape of an elbow. From the onset of this pattern, one can decide which components to keep to do a dimensionality reduction since eigenvalues above this point do not contribute much to explain the remaining variance. This, however, is very imprecise and subjective.

Another way to decide what components to keep is by calculating the cumulative proportion of explained variance (CPEV):

$$\sigma_{n,ac}^2 = \sum_{i=1}^{n} \sigma_{i,\%}^2 = \frac{\sum_{i=1}^{n} \lambda_i}{\sum_m \lambda_m}, \tag{3.9}$$

where the summation index is over the descending-ordered eigenvalues. Reading the plot from left to right it is possible to decide the number of components to keep by checking when a specific CPEV has been obtained.

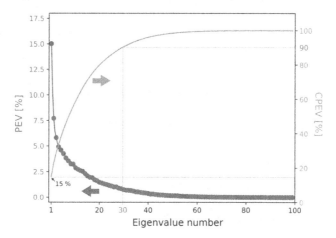

Figure 3.2   The PEV and CPEV plots as functions of the eigenvalue number

In Julia, we can simply do:

```
s = sum(va)
PEV = va./s
CPEV = cumsum(va)./s
```

An example from random data is shown in Fig. 3.2. The first eigenvalue contributes 15 % of the PEV and this is the onset of the CPEV curve. After reaching 90 % of the CPEV, for example, one can decide that this is sufficient and retain the corresponding eigenstates. In this example, this is achieved with the first 30 eigenvalues.

3.1.3   PROJECTION ON NEW BASIS

Eigenvalues that contribute less than a determined threshold to the explained variance are typically discarded together with their respective eigenvectors because they do not help much in explaining the variability found in the complete dataset. The discarded eigenstates are typically considered noise. Therefore, this operation is often described as a way of *denoising* the data.

In our example, we will keep only the eigenvector whose corresponding cumulated explained variance is higher than 10 % and construct a weight matrix $\mathbf{W} \in \mathbb{R}^{b \times m}$, where D is the new dimensionality of the system (in this case $D = 2$):

$$\mathbf{W} = \begin{bmatrix} \mathbf{v}_3 & \mathbf{v}_1 \end{bmatrix}. \tag{3.10}$$

The complete dataset can now undergo a dimensionality reduction by being projected onto this new system $\mathbf{W}$:

$$\mathbf{R}' = \mathbf{W}^\top \mathbf{R}, \tag{3.11}$$

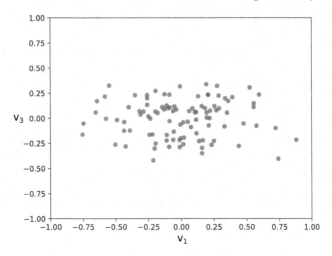

Figure 3.3   Data from Fig. 3.1 after dimensionality reduction and projected on axes v_1 and v_3

where $\mathbf{R}' \in \mathbb{R}^{D \times N}$. The data from Fig. 3.1 under a dimensionality reduction is shown in Fig. 3.3.

In Julia:

```
W = hcat(ve[:,1],ve[:,2])
R = W'*R
scatter(R[1,:],R[2,:])
```

3.2 SINGULAR VALUE DECOMPOSITION

Another way to compute the principal components is using the *singular value decomposition* (SVD) developed by Jordan[d] and Beltrami[e] [3, 4]. This comprises of decomposing an $m \times n$ matrix $\mathbf{X}$ as:

$$\mathbf{X} = \mathbf{U}\Sigma\mathbf{V}^\top, \tag{3.12}$$

where $\mathbf{U}$ and $\mathbf{V}^\top$ are $m \times m$ and $n \times n$ unitary matrices[f] (real or complex), respectively, whose columns of each constitute orthogonal basis vectors. The columns of $\mathbf{U}$ and $\mathbf{V}$ are called, respectively, the *left and right singular vectors* of $\mathbf{X}$ The matrix Σ is a $m \times n$ diagonal matrix whose (singular) elements are non-negative real numbers. The number of non-zero singular elements gives the rank of the matrix $\mathbf{X}$. Treating each

[d]Marie Ennemond Camille Jordan (1838–1922) French mathematician.
[e]Eugenio Beltrami (1825–1900) Italian mathematician.
[f]$\mathbf{U}^\top\mathbf{U} = \mathbf{V}^\top\mathbf{V} = \mathbf{I}$.

matrix as a linear transformation, the decomposition can be understood as a rotation (or reflection) given by $\mathbf{V}^\top$ followed by a scaling given by Σ and another rotation (or reflection) given by $\mathbf{U}$. This is illustrated in Fig. 3.4.

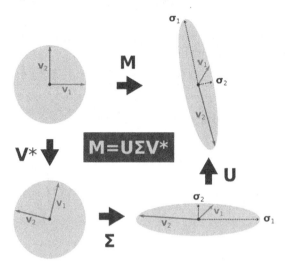

Figure 3.4 The SVD of the linear map $\mathbf{M} = \begin{bmatrix} 0.30 & 0.45 \\ 0.50 & -1.6 \end{bmatrix}$ applied to two orthogonal vectors $\mathbf{v}_{1,2}$. $\sigma_{1,2}$ are the singular values of $\mathbf{M}$

For a real matrix $\mathbf{X}$, we can write:

$$\begin{aligned} \mathbf{X}^\top \mathbf{X} &= \mathbf{V}\Sigma \mathbf{U}^\top \mathbf{U}\Sigma \mathbf{V}^\top \\ &= \mathbf{V}\left(\Sigma \odot \Sigma\right)\mathbf{V}^\top \\ &= \mathbf{V}\Sigma^2 \mathbf{V}^\top. \end{aligned} \tag{3.13}$$

This is exactly the eigendecomposition of the correlation matrix $\mathbf{X}^\top \mathbf{X}$. Therefore, $\mathbf{V}$ is formed with the eigenvectors of the correlation matrix and the singular values are the square root of its eigenvalues.

Example 3.2.1. SVD

For a data matrix:

$$\mathbf{X} = \begin{bmatrix} 1 & 2 \\ 3 & 4 \\ 5 & 6 \end{bmatrix}, \tag{3.14}$$

we calculate:

$$\mathbf{C} = \mathbf{X}^\top \mathbf{X} = \begin{bmatrix} 35 & 44 \\ 44 & 56 \end{bmatrix}. \tag{3.15}$$

This gives an eigendecomposition $C = Q\Lambda Q^\top$:

$$\Lambda = \begin{bmatrix} 90.73 & \\ & 0.26 \end{bmatrix}, \quad Q = \begin{bmatrix} -0.62 & -0.78 \\ -0.78 & 0.62 \end{bmatrix}. \tag{3.16}$$

The singular values are the square-root values of these eigenvalues and $V = Q$. Now, to find U we work with Eq. 3.12:

$$U = X \left(\Sigma V^\top \right)^{-1}. \tag{3.17}$$

As a result, we end up with:

$$
\begin{array}{cccc}
X & U & \Sigma & V^\top
\end{array}
$$

$$
\begin{bmatrix} 1 & 2 \\ 3 & 4 \\ 5 & 6 \end{bmatrix} = \begin{bmatrix} -0.23 & 0.99 & 0.41 \\ -0.52 & 0.24 & -0.82 \\ -0.82 & -0.40 & 0.41 \end{bmatrix} \begin{bmatrix} 9.52 & \\ & 0.51 \\ & \end{bmatrix} \begin{bmatrix} -0.62 & -0.78 \\ -0.78 & 0.62 \end{bmatrix}.
$$

$$\tag{3.18}$$

The projection on the principal components (right singular vectors) is given by:

$$X' = XV = U\Sigma V^\top V = U\Sigma, \tag{3.19}$$

which is the left singular vectors scaled by the singular values of X. This is also called the *polar decomposition* of X'.

The advantage of using SVD is that numerous fast and stable methods exist, which do not depend on the direct computation of the correlation matrix. Examples include the power method[g] [34] and the Lanczos[h] algorithm [35].

3.2.1 PSEUDOINVERSE

Singular value decomposition can also be used to simplify the calculation of the pseudoinverse. Consider a linear system:

Applying SVD to matrix $A = U\Sigma V^\top$:

$$
\begin{aligned}
A^+ &= \left(V\Sigma^2 V^\top \right)^{-1} V\Sigma U^\top \\
&= V\Sigma^{-2} V^\top V\Sigma U^\top \\
&= V\Sigma^{-1} U^\top.
\end{aligned}
\tag{3.20}
$$

This is rather simple to compute. Note, however, that all diagonal elements of Σ must be non-zero.

---

[g] Developed by Richard von Mises (1883–1953) Austrian engineer and mathematician younger brother of the economist Ludwig von Mises.

[h] Cornelius Lanczos (1893–1974) Hungarian mathematician.

3.3   KERNEL PCA

So far we have assumed that our data is linearly separable as in the example of Fig. 3.1. Trying to apply the PCA approach to non-linearly separable data often produces meaningless results. In this section, we will deal with an extension of the PCA technique applied to the non-linear case and we will see that it can be useful for data classification.

Consider, for example, the data set shown in Fig. 3.5a. It is clearly non-linearly separable, but if we add another dimension $z_i = x_i^2 + y_i^2$, we get the separable set shown in Fig. 3.5b.

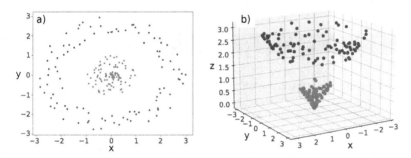

Figure 3.5   a) Non-linearly separable data on an x-y plane. b) Same dataset projected on a higher dimension

Applying a transformation and then performing a regular PCA can be used to classify the data, but it is often computationally intense, especially if the dataset has many dimensions. Instead, we will use a *kernel* approach [36].

For this, we take the column vectors $\mathbf{r}_i$ to a higher dimensional (feature) space by applying a function $\phi : \mathbb{R}^d \to \mathbb{R}^D$. This can be applied to the whole data matrix $\mathbf{R}$ using a function $\Phi : \mathbb{R}^{d \times N} \to \mathbb{R}^{D \times N}$. The corresponding covariance matrix $\mathbf{C} \in \mathbb{R}^{D \times D}$ is given by:

$$\mathbf{C} = \Phi(\mathbf{R})\Phi(\mathbf{R})^\top, \tag{3.21}$$

and its eigenstate equation is:

$$\mathbf{CV} = \mathbf{V}\Lambda$$
$$\Phi(\mathbf{R})\Phi(\mathbf{R})^\top \mathbf{V} = \mathbf{V}\Lambda, \tag{3.22}$$

where $\mathbf{V} \in \mathbb{R}^{D \times D}$ is a matrix whose columns are the eigenvectors of the correlation matrix, and $\Lambda \in \mathbb{R}^{D \times D}$ is a diagonal matrix with the eigenvalues of the correlation matrix.

Let's now write the eigenvectors as linear combinations of the transformed structures:

$$\mathbf{V} = \Phi(\mathbf{R})\mathbf{A}, \tag{3.23}$$

where $\mathbf{A} \in \mathbb{R}^{N \times D}$.

Consequently, Eq. 3.22 becomes:

$$\Phi(\mathbf{R})\Phi(\mathbf{R})^{\top}\Phi(\mathbf{R})\mathbf{A} = \Phi(\mathbf{R})\mathbf{A}\Lambda \tag{3.24}$$

Multiplying the whole expression by $\Phi(\mathbf{R})$ from the left:

$$\Phi(\mathbf{R})^{\top}\Phi(\mathbf{R})\Phi(\mathbf{R})^{\top}\Phi(\mathbf{R})\mathbf{A} = \Phi(\mathbf{R})^{\top}\Phi(\mathbf{R})\mathbf{A}\Lambda \tag{3.25}$$

Now, instead of explicitly computing the products using the unknown function Φ, we use a kernel function $\mathbf{K}: \mathbb{R}^{D \times N} \to \mathbb{R}^{N \times N}$ such that $\mathbf{K}(\mathbf{R}) = \Phi(\mathbf{R})^{\top}\Phi(\mathbf{R})$:

$$\mathbf{KA} = \mathbf{A}\Lambda. \tag{3.26}$$

Mercer's Kernels

For a nonempty index set χ, a symmetric function $K: \chi \times \chi \to \mathbb{R}$ is known as the (positive-definite) kernel on χ if:

$$\sum_{i,j=1} c_i c_j K(x_i, x_j) \geq 0, \ \forall x_i \in \chi, \ c_i \in \mathbb{R}. \tag{3.27}$$

Given a specific positive definite kernel, Mercer's[a] theorem [37] guarantees the existence of a function Φ if and only if $K(\mathbf{x},\mathbf{y}) = K(\mathbf{y},\mathbf{x})$ and $\iint K(\mathbf{x},\mathbf{y})f(\mathbf{x})f(\mathbf{y})d\mathbf{x}d\mathbf{y} \geq 0$, $f \in L^2$. The table below shows some of the most common kernels used in machine learning.

Kernel	$K(\mathbf{x},\mathbf{y})$
Linear	$\mathbf{x} \cdot \mathbf{y}$
Polynomial	$(\mathbf{x} \cdot \mathbf{y} + r)^n$, $r \geq 0$, $n \geq 1$
Gaussian	$\exp\left(-\gamma\|\mathbf{x}-\mathbf{y}\|^2\right)$, $\gamma > 0$
Laplacian	$\exp\left(-\alpha\|\mathbf{x}-\mathbf{y}\|\right)$, $\alpha > 0$
Hyperbolic tangent	$\tanh(\mathbf{x} \cdot \mathbf{y} + \delta)$.

---

[a]James Mercer (1883–1932) English mathematician.

Note, however, that the data matrix has to be centered. For this, we make:

$$\hat{\Phi}(\mathbf{R}) = \Phi(\mathbf{R}) - \frac{1}{N}\Phi(\mathbf{R})\mathbf{J}_N = \Phi(\mathbf{R})\left(\mathbf{I}_N - \frac{1}{N}\mathbf{J}_N\right), \tag{3.28}$$

where $\mathbf{J}_N$ is a $N \times N$ unit matrix[i] whose all elements are 1 and $\mathbf{I}_N$ is an $N \times N$ identity matrix.

---

[i]Not be confused with the unitary and identity matrices.

The kernel then becomes:

$$\hat{\mathbf{K}} = \hat{\Phi}(\mathbf{R})^\top \hat{\Phi}(\mathbf{R})$$

$$= \left(\mathbf{I}_N - \frac{1}{N}\mathbf{J}_m\right)\Phi(\mathbf{R})^\top\Phi(\mathbf{R})\left(\mathbf{I}_N - \frac{1}{N}\mathbf{J}_N\right)$$

$$= \left(\mathbf{I}_N - \frac{1}{N}\mathbf{J}_N\right)\mathbf{K}\left(\mathbf{I}_N - \frac{1}{N}\mathbf{J}_N\right). \tag{3.29}$$

In Julia, we write:

```
using LinearAlgebra

K = [kernel(X[:,i],X[:,j]) for i in 1:N, j in 1:N]

J = ones(N,N)
Kp = (I − J/N)*K*(I − J/N)
```

The eigen equation 3.26 becomes:

$$\hat{\mathbf{K}}\mathbf{A} = \mathbf{A}\Lambda. \tag{3.30}$$

This gives us the eigenstates of the kernel, but we actually need the eigenstates of the correlation matrix. To find them, we use the SVD for $\Phi(\mathbf{R}) = \mathbf{U}\Sigma\mathbf{V}^\top$. Therefore, we have for the correlation matrix:

$$\mathbf{C} = \mathbf{U}\Sigma^2\mathbf{U}^\top. \tag{3.31}$$

And for the kernel matrix, we have:

$$\hat{\mathbf{K}} = \mathbf{V}\Sigma^2\mathbf{V}^\top. \tag{3.32}$$

We see that both share the same eigenvalues, but have different eigenvectors. Also, $\mathbf{V} = \mathbf{A}$ and $\Lambda = \Sigma^2$.

We can find the eigenstates of the kernel in Julia by doing:

```
Σ = Diagonal(sqrt.(eigvals(Kp)))
A = eigvecs(Kp)

A = reverse(A,dims=2)
v = reverse(v) v[v . < 0] .= 0
Σ = Diagonal(sqrt.(v))
```

Note that the obtained eigenvalues are Σ^2 and we used the *sqrt* function to obtain Σ. Also, we sorted the eigenstates from the biggest to the smallest and discarded meaningless eigenvalues.

To find the projection of the dataset on the principal axes, we compute:

$$\Phi(\mathbf{R})' = \mathbf{U}^\top \Phi(\mathbf{R})$$
$$= (\mathbf{V}\Sigma)^{-1} \Phi(\mathbf{R})^\top \Phi(\mathbf{R}). \tag{3.33}$$

One can easily recognize the kernel matrix in the last expression, but it can be written using the result from Eq. 3.32:

$$\Phi(\mathbf{R})' = (\mathbf{V}\Sigma)^{-1} \mathbf{V}\Sigma^2 \mathbf{V}^\top$$
$$= \Sigma \mathbf{V}^\top. \tag{3.34}$$

In Julia:

```
nΦ = Σ*A'
```

Figure 3.6a shows a non-linearly separable dataset, whereas Fig. 3.6b shows the projection on the kernel-PCA axes. The projection is clearly separable and classifiable.

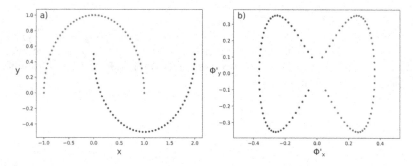

Figure 3.6 a) Non-linearly separable dataset and b) its projection on the kernel-PCA axes. A Gaussian kernel with $\gamma = 15$ was used

3.4 INDEPENDENT COMPONENT ANALYSIS

Imagine a scenario where there are n statistically independent audio sources and m microphones in the same room. Audio is collected during a period T such that the sources can be grouped into a matrix $\mathbf{S} \in \mathbb{R}^{n \times T}$ and the microphones into a matrix $\mathbf{R} \in \mathbb{R}^{m \times T}$. The latter is, actually, a mixture of the sources:

$$\mathbf{R} = \mathbf{AS}, \tag{3.35}$$

where $\mathbf{A} \in \mathbb{R}^{n \times m}$ is known as the *mixing matrix*, which is typically unknown.

Our problem is known as *blind source separation* (BSS) [38, 39]. Given the matrix **R**, how to estimate each source individually? Mathematically:

$$\hat{\mathbf{S}} = \mathbf{WR} = \mathbf{WAS}, \qquad (3.36)$$

where $\mathbf{W} \in \mathbb{R}^{m \times n}$ is called *unmixing matrix*, which we need to estimate. To perform this operation, first, we will *sphere* the data (a procedure similar to the one we did in Sec. 3.1). This corresponds to centering it:

$$\text{row}_i \mathbf{R} \mapsto \text{row}_i \mathbf{R} - \langle \text{row}_i \mathbf{R} \rangle . \qquad (3.37)$$

In Julia:

```julia
using Statistics
using LinearAlgebra

# Centering
R = (R .- mean(R,dims=2))
```

Next, we transform the data so that its covariance is unitary so that we eliminate correlations (*whitening*). For this, we look for a linear map $\mathbf{T} \in \mathbb{R}^{n \times n}$ such that the whitened data becomes:

$$\mathbf{R}_w = \mathbf{TR}. \qquad (3.38)$$

To make the covariance unitary, we must do:

$$\mathbf{R}_w \mathbf{R}_w^\top = \mathbf{T} \left(\mathbf{R}\mathbf{R}^\top \right) \mathbf{T}^\top$$
$$\mathbf{I} = \mathbf{TCT}^\top \qquad (3.39)$$
$$\mathbf{T}^\top \mathbf{T} = \mathbf{C}^{-1}.$$

This last matrix $\mathbf{C} \in \mathbb{R}^{n \times n}$ can be eigen-decomposed (see Eq. 3.31) as:

$$\mathbf{C} = \mathbf{U}\Sigma^2\mathbf{U}^\top$$
$$= \mathbf{U}\Sigma\Sigma\mathbf{U}^\top \qquad (3.40)$$
$$\mathbf{C}^{-1} = \left(\mathbf{U}\Sigma^{-1} \right) \left(\Sigma^{-1}\mathbf{U}^\top \right) = \mathbf{T}^\top\mathbf{T}.$$

Therefore, a good option for the transformation matrix is $\mathbf{T} = \Sigma^{-1}\mathbf{U}^\top$. One inconvenience of this transformation is that it rotates and reflects the data as we have previously seen in Sec. 3.2 and this destroys the spatial correlation of the data[j]. This

---

[j]Given the analogy, this is often referred to as PCA whitening.

is a problem, for example, if we are dealing with images. One simple way to circumvent this problem is to rotate the data back to its original orientation by applying the rotation matrix $\mathbf{U}$ from the left[k] such that $\mathbf{T} = \mathbf{U}\mathbf{\Sigma}^{-1}\mathbf{U}^\top$, which is just $\mathbf{C}^{-1/2}$.

On the other hand, we want to find the axes of maximum independence. Note that the whitening procedure already removes second-order correlations. If instead of applying the rotation matrix $\mathbf{U}$ from the left, we use another rotation matrix $\mathbf{V} \in \mathbb{R}^{n \times m}$ that removes higher-order correlations, the independent component analysis (ICA) problem is solved. Therefore, we can write $\hat{\mathbf{S}} = \mathbf{V}\mathbf{R}_w$.

The sphering operation in Julia is illustrated below:

```
# ZCA Whitening
C = R*R'
v = eigvals(C)
U = eigvecs(C)

Σ = Diagonal(1 ./ sqrt.(v))

T = Σ*U'
Rw = T*R
```

To estimate matrix $\mathbf{V}$, we will use a *contrast function* [41] $\mathbb{R}^n \to \mathbb{R}$ as an estimate of non-Gaussianity. In our case, let's use the average value of a function $g : \mathbb{R} \to \mathbb{R}$ evaluated on every element of a vector. Let's now build a Lagrange function using the constraint that $\mathbf{v}^\top\mathbf{v} = 1$ since $\mathbf{V}$ is a rotation matrix:

$$\Lambda = \left\langle g(\mathbf{v}^\top\mathbf{R}_w) \right\rangle - \lambda(\mathbf{v}^\top\mathbf{v} - 1). \tag{3.41}$$

Here the average is over the time series, the resulting elements of the row vector. Therefore, we can write: $\langle \mathbf{u}^\top \rangle = \frac{1}{T}\mathbf{u}^\top\mathbf{1}$.

Minimizing this Lagrange function, we obtain the multiplier:

$$h(\mathbf{v}^\top\mathbf{R}_w)\mathbf{R}_w^\top - \lambda\mathbf{v}^\top = 0 \to \lambda = h(\mathbf{v}^\top\mathbf{R}_w)\left(\mathbf{v}^\top\mathbf{R}_w\right)^\top. \tag{3.42}$$

where $h = g'$.

---

[k]This is also known as *zero-phase component analysis* (ZCA) [40] and is also sometimes referred to as Mahalonobis transformation, after Prasanta Chandra Mahalonobis (1893–1972) adviser of Calyampudi Radhakrishna Rao, among others.

Newton-Raphson Method

The best linear approximation for a differentiable function $\mathbf{f} : \mathbb{R}^n \to \mathbb{R}^m$ around a point $\mathbf{v}^{(k)}$ is given by:

$$\mathbf{f}\left(\mathbf{v}^{(k+1)}\right) \approx \mathbf{f}\left(\mathbf{v}^{(k)}\right) + \mathbf{J}_f\left(\mathbf{v}^{(k+1)} - \mathbf{v}^{(k)}\right), \qquad (3.43)$$

where $\mathbf{J}_f = \frac{\partial \mathbf{f}}{\partial \mathbf{v}} \in \mathbb{R}^{m \times n}$ is the Jacobian[a] matrix.

To find the roots of $\mathbf{f}$, we want the function at the next point $\mathbf{v}_{k+1}$ to be zero. Therefore:

$$\mathbf{v}^{(k+1)} \approx \mathbf{v}^{(k)} - \mathbf{J}_f^{-1}\mathbf{f}\left(\mathbf{v}^{(k)}\right). \qquad (3.44)$$

This is known as the Newton-Raphson[b][c] method for optimization [3].

---

[a]Carl Gustav Jacob Jacobi (1804–1851) German mathematician.
[b]Isaac Newton (1642–1726) English natural philosopher and Joseph Raphson (1648–1715) English mathematician.
[c]See Sec. 7.1 for a second-order version.

To find $\mathbf{v}$ from the last equation, we can use Newton's method for the function:

$$\mathbf{f} = \mathbf{R}_w h(\mathbf{R}_w^\top \mathbf{v}) - \lambda \mathbf{v} \qquad (3.45)$$

and compute its Jacobian:

$$\mathbf{J} = \mathbf{R}_w \operatorname{diag}\left[l(\mathbf{v}_1^\top \mathbf{R}_w)\right] \mathbf{R}_w^\top - \lambda \mathbf{I}, \qquad (3.46)$$

where $l = h'$.

Thus, we arrive at the following independent component analysis (ICA) algorithm [38, 39]:

Algorithm 4: ICA

---

input: Independent units
$\mathbf{v}$ = random vector;
repeat
  $\triangleright\ \lambda = \left(\mathbf{v}^\top \mathbf{R}_w\right) \cdot h(\mathbf{v}^\top \mathbf{R}_w)$;
  $\triangleright\ \mathbf{f} = \mathbf{R}_w h(\mathbf{R}_w^\top \mathbf{v}) - \lambda \mathbf{v}$;
  $\triangleright\ \mathbf{J} = \mathbf{R}_w \operatorname{diag}\left[l(\mathbf{v}_1^\top \mathbf{R}_w)\right] \mathbf{R}_w^\top - \lambda \mathbf{I}$;
  $\triangleright\ \mathbf{v} \mapsto \mathbf{v} - \eta \mathbf{J}^{-1}\mathbf{f}$;
  $\triangleright\ \mathbf{v} \mapsto \frac{\mathbf{v}}{\|\mathbf{v}\|}$;
until *convergence*;

A typical choice for the contrast function is the kurtosis, such that $g(u) = u^4$, $h(u) = 4u^3$, and $l(u) = 12u^2$.

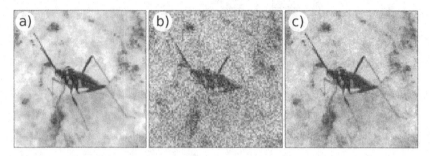

Figure 3.7   a) An original image, b) this image mixed with uncorrelated noise, and c) estimated image using ICA

This algorithm implemented in Julia becomes:

```
while (error < threshold)
    t = v'*Rw
    λ = (h.(t)*t')[1,1]
    f = Rw*h.(t') − λ*v
    J = Rw*diagm(l.(t)[1,:])*Rw' − λ*I
    error = inv(J)*f
    p = v − η*error
    v = p/norm(p)
end
```

To find the other vectors, different initial values for **v** are used and then a Gram-Schmidt[1] decorrelation [3] procedure can be used at every step of the ICA algorithm:

$$\mathbf{v}_k \mapsto \mathbf{v}_k - \sum_{i=1}^{k-1} \frac{\mathbf{v}_i \cdot \mathbf{v}_k}{\|\mathbf{v}_i\|} \mathbf{v}_i. \tag{3.47}$$

Here the idea is to remove the projection of the previous vectors from the current one so that they become orthogonal.

In Fig. 3.7, a picture is mixed with uncorrelated noise and then estimated using ICA.

---

[1]Jørgen Pedersen Gram (1850–1916) Danish mathematician and Erhard Schmidt (1876–1959) Estonian mathematician, advised by David Hilbert.

4 Cluster Analysis

Objects that share similar features within a dataset form a cluster. Clustering objects helps us discover natural groups that exist in the data, find patterns and trends, and segment images into meaningful regions, for example.

In this chapter, we will discuss four groups of clustering algorithms: K-means and K-medoids are simple algorithms that partition the data into K clusters. DBSCAN (density-based spatial clustering of applications with noise) is a more sophisticated algorithm based on density that can handle noise more efficiently. Hierarchical algorithms work by constructing a hierarchy of clusters by merging or splitting clusters. When data cannot be easily separated by traditional distance-based methods, clustering the data according to the eigen-components of a corresponding graph matrix can be used to group similar features together.

4.1 K-MEANS AND K-MEDOIDS

The k-means [42] and k-medoids [43] algorithms have the main objective of partitioning a data set $\mathbb{X} = \{\mathbf{x}_1, \ldots, \mathbf{x}_N\}$ into a cluster set $\mathbb{C} = \{C_1, \ldots, C_K\}$. For k-means, we define centroids μ_k corresponding to these clusters and a function $v : \mathbb{X}, \mathbb{C} \to \{0, 1\}$ such that $v(n, k) = 1$ if data $\mathbf{x}_n \in C_k$ and 0 otherwise.

Also, we define an objective function:

$$J = \sum_{n=1}^{N} \sum_{k=1}^{K} v(n, k) \|\mathbf{x}_n - \mu_k\|^2 \tag{4.1}$$

that describes the cost of the data being segmented in a current cluster configuration.

The k-means algorithm works by repeating two steps until convergence. In the first step, data are assigned to clusters such that:

$$v(n, k) = \begin{cases} 1 & \text{if } k = \arg \min_j \|\mathbf{x}_n - \mu_j\|^2 \\ 0 & \text{otherwise.} \end{cases} \tag{4.2}$$

In Julia, we can organize the data in an array where the first column is the x-values, the second is the y-values, and the third is the cluster to which the data point belongs:

DOI: 10.1201/9781003350101-4

```
for i in 1:N          # over data
    s_min=1E5
    for k in 1:3       # over clusters
        s=sum((R[i,1:2]-C[k,>]).^2)
        if (s<s_min)
            s_min=s
            R[i,3]=k
        end
    end
end
```

In the second step, we find the centroids that minimize the objective function:

$$\frac{\partial J}{\partial \mu_p} = \sum_{n=1}^{N} v(n,p)(x_n - \mu_p) = 0 \rightarrow \mu_p = \frac{\sum_{n=1}^{N} v(n,p)x_n}{\sum_{n=1}^{N} v(n,p)}, \qquad (4.3)$$

which is equivalent to:

$$\mu_p = \frac{1}{|C_p|} \sum_{r \in C_p} r. \qquad (4.4)$$

In Julia:

```
C = zeros(3,2)
num = zeros(3)
for i in 1:N
    p = trunc(Int,R[i,3])
    C[p,:] = C[p,:] + R[i,1:2]
    num[p] = num[p] + 1
end
for i in 1:3
    C[i,:] = C[i,:]/num[i]
end
```

Similar to a centroid, a *medoid* is defined as one element from this data set whose dissimilarity to all other elements is the smallest. Mathematically, this is described by:

$$x_m = \arg \min_{y \in \mathbb{X}} \sum_{i=1}^{N} d(x_i, y), \qquad (4.5)$$

where d is the dissimilarity measure.

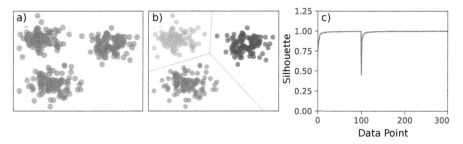

Figure 4.1   a) raw data, b) clustered data, and c) silhouette function

For k-medoids, one usually uses the *partitioning around medoids* (PAM) algorithm [43] which is very similar to the k-means algorithm. The main difference is the update step where new elements are tested as medoids. If a related cost function is reduced, then this data point becomes the new medoid. Typically, the average intracluster dissimilarity is used as a cost function. In Julia:

```
BestCost = Cost(Medoids,R)
for i in 1:N
    MedoidsBK = Medoids
    p = trunc(Int,R[i,3])
    Medoids[p] = i
    c = Cost(Medoids,R)
    if (c < BestCost)
        BestCost = c
    else
        Medoids = MedoidsBK
    end
end
```

4.1.1 SILHOUETTE

We can check the quality of the produced clusters with the *silhouette* [44]. This gives an indication of the similarity between an element and the cluster to which it belongs. A value of $+1$ indicates a good match, whereas a value of -1 indicates that the element is separated from the group.

To calculate the silhouette, first, we need to compute how well a point $\mathbf{x}_n$ is adjusted to its cluster C:

$$a(\mathbf{x}) = \frac{1}{|C| - 1} \sum_{(\mathbf{y} \neq \mathbf{x}) \in C} d(\mathbf{x}, \mathbf{y}). \tag{4.6}$$

This is normalized by the minimum dissimilarity between this point and the other clusters:

$$b(\mathbf{x}) = \min_{S \neq C} \frac{1}{|S|} \sum_{y \in S} d(\mathbf{x}, \mathbf{y}). \tag{4.7}$$

The silhouette value is then computed for each point within a cluster as:

$$s(\mathbf{x}) = \begin{cases} 1 - a(\mathbf{x})/b(\mathbf{x}), & \text{if } a(\mathbf{x}) < b(\mathbf{x}) \\ 0, & \text{if } a(\mathbf{x}) = b(\mathbf{x}) \\ b(\mathbf{x})/a(\mathbf{x}) - 1, & \text{if } a(\mathbf{x}) > b(\mathbf{x}). \end{cases} \tag{4.8}$$

In Julia, to find a and b, for each element we must compute the dissimilarities:

```
t = zeros(NC)
p = trunc(Int,R[i,3])
a = 0
for j in 1:N
    q = trunc(Int,R[j,3])
    d = Euclidean(R[i,1:2],R[j,1:2])
    if (p == q)
        a = a + d/(nC[q]-1)
    else
        t[q] = t[q] + d/nC[q]
    end
end
b = minimum(t[1:NC .≠ p])
```

It is possible to use the silhouette function to determine the number of clusters to use in k-means and k-medoids. For this, we compute an average silhouette function and find the value of k that maximizes it.

4.2 DBSCAN

One problem of the k-means and k-medoids algorithms is that one needs to establish a fixed number of clusters before running them. Furthermore, the clusters need to be linearly separable. The DBSCAN [45] circumvents both problems.

This algorithm checks if a point has a sufficient number of neighbors (N_{min}) by calculating the distance between them ($d < d_{min}$). If so, then it is classified as a *core point* belonging to a cluster, and other *directly reachable points* are recursively added to this cluster. All points in the cluster will be *reachable* to the original core point through a path. If the original point has an insufficient number of neighbors, then it is considered an *outlier* or noise.

Algorithm 5: DBSCAN

---

input: Set of points
▷ $C = 0$;
for *each point* **p** *in set* **do**
  if *cluster*(**p**) *is undefined* **and** *neighborhood*(**p**) $> N_{\min}$ **then**
    │ ▷ $C = C + 1$;
    │ ▷ Clusterize **p** in C;
  end
end
Function *clusterize(***p***,cluster)* **is**
  │ ▷ Create *list* with element **p**;
  │ **while** *list is not empty* **do**
  │   │ ▷ **q** ← pop from *list*;
  │   │ **if** *cluster*(**q**) *is undefined* **and** *neighborhood*(q) $> N_{\min}$ **then**
  │   │   │ ▷ Point **q** now belongs to *cluster*;
  │   │   │ ▷ Push each neighbor of **q** into *list*;
  │   │ **end**
  │ **end**
end
Function *neighborhood(***p***)* **is**
  │ **return** *All points* **q** *such that* $\|\mathbf{p} - \mathbf{q}\| < d_{\min}$
end

Some advantages of DBSCAN over k-means are that i) it works with non-linearly separable data as illustrated in Fig. 4.2, and ii) it is robust to outliers.

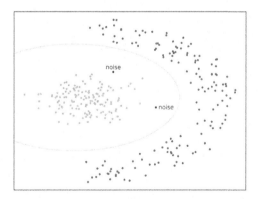

Figure 4.2   Non-linearly separable data clustered with the DBSCAN algorithm

4.3   HIERARCHICAL CLUSTERING

Hierarchical clustering aims at constructing a binary tree (dendrogram) that represents the taxonomy of relationships between groups of units. The tree is composed of stacking branches (clades) that recursively break into smaller branches, and terminates with units (leaves). The dendrogram can be formed either in a bottom-up strategy where units are progressively grouped together or in a top-down approach where all units began as a single cluster and then it is progressively broken down. The first strategy is known as *agglomerative nesting* (AGNES) whereas the second one is known as *divisive analysis* (DIANA) [46].

Figure 4.3 shows a dendrogram produced using AGNES for the logarithmic returns [24] of seven stocks (IBM, INTC, NVDA, AMD, TXN, ADI, and MCHP) obtained from Yahoo Finance for a period of 254 days beginning on 2022/06/22.

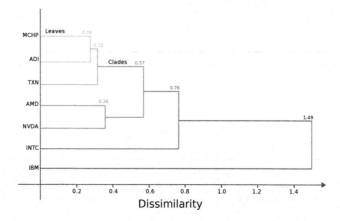

Figure 4.3   Dendrogram using AGNES for the log returns of seven stocks obtained from Yahoo Finance for a period of 254 days beginning on 2022/06/22. Dissimilarity was estimated as the average Euclidean distance, and a simple average clustering strategy was used

To cluster the elements together, we need a metric that allows us to quantify how similar they are. One typical generalized metric is the Minkowski[a] distance:

$$d(\mathbf{x}, \mathbf{y}) = \left(\sum_{i=1}^{N} \|x_i - y_i\|^p \right)^{1/p} , \quad p \in \mathbb{Z}. \tag{4.9}$$

This metric reduces to the Euclidean[b] distance when $p = 2$ and to the Manhattan distance when $p = 1$.

---

[a]Hermann Minkowski (1864–1909) German mathematician.
[b]Euclid of Alexandria (300 BC) Greek mathematician.

4.3.1 AGNES

The AGNES clustering strategy is described by the following algorithm [46, 47]:

Algorithm 6: AGNES

---

input: N independent units
$\triangleright$ Create a proximity matrix $\mathbf{D} \in \mathbb{R}_{\geq 0}^{N \times N}$ with all distances $d(\mathbf{e}_i, \mathbf{e}_j)$;
repeat
$\quad \triangleright$ Find the most similar pair with elements $\mathbf{e}_a$ and $\mathbf{e}_b$;
$\quad \triangleright$ Remove $\mathrm{col}_{a,b}\mathbf{D}$ and $\mathrm{row}_{a,b}\mathbf{D}$;
$\quad \triangleright$ Include a new element $\mathbf{e}_c$ corresponding to the clustered pair;
$\quad \triangleright$ Update $\mathbf{D}$ with distances $d(\mathbf{e}_c, \mathbf{e}_k) =$
$\quad \alpha_a d(\mathbf{e}_a, \mathbf{e}_k) + \alpha_b d(\mathbf{e}_b, \mathbf{e}_k) + \beta d(\mathbf{e}_a, \mathbf{e}_b) + \gamma |d(\mathbf{e}_a \mathbf{e}_k) - d(\mathbf{e}_b, \mathbf{e}_k)|$;
until *all units are clustered*;

Depending on the parameters $\alpha_{a,b}$, β, and γ, different techniques can be conceived. For instance, if $\alpha_a = \alpha_b = 0.5$, $\beta = 0$, and $\gamma = -0.5$ we have the *single-linkage* or *nearest neighbor clustering* [46, 47]. Note that this corresponds to the minimum distance between elements a and b. Another option is $\alpha_a = \alpha_b = 0.5$, $\beta = \gamma = 0$, which corresponds to a *simple average clustering* [46–48]. In all possibilities, the new distance is smaller than the maximum of the previous distances. Therefore, hierarchical clustering produces an ultrametric ordered graph.

In Julia, we can use the fact that the proximity matrix is symmetric. If *NClu* is the number of elements, we can find the most similar pair (c_1, c_2) with:

```
for i in 1:(NClu-1)
    for j in (i+1):NClu
        if (D[i,j] < d_min)
            d_min = D[i,j]
            c1 = i
            c2 = j
        end
    end
end
```

Now, we can create a map of the elements of the proximity matrix to a new one eliminating c_1 and c_2:

```
map = zeros(Int,NClu)
j = 1

for i in 1:NClu
    if (i != c1 && i != c2)
        map[i] = j
        j = j + 1
    else
        map[i] = 0
    end
end
```

Next, we copy the elements of the proximity matrix D to the new one nD. Also, we include a variable *Label* and a new label nL to keep track of the units being clustered together:

```
nL = ["" for i in 1:NClu−1]
nD = zeros(NClu-1,NClu−1)

for j in 1:NClu
    if (j != c1 && j != c2)
        nL[map[j]] = Labels[j]
        for i in 1:NClu
            if (i != c1 && i != c2)
                nD[map[i],map[j]] = D[i,j]
            end
        end
    end
end
```

Finally, we add the new distances according to one of the strategies and also include a new label:

```
for i in 1:NClu
    if (i != c1 && i != c2)
        nL[NClu-1] = Labels[c1]*"/"*Labels[c2]
        d = 0.5*(D[i,c1]+D[i,c2])
        nD[map[i],NClu-1] = d
        nD[NClu-1,map[i]] = d
    end
end

NClu = NClu - 1
Labels = nL
D = nD
```

4.3.2 DIANA

For doing a top-down divisive clustering [46], we begin with a single cluster C, and consider two new ones A and B such that, initially, $A = C$ and $B = \emptyset$. Then, for each element $\mathbf{x} \in A$ we compute the average dissimilarities to all other elements of A:

$$\bar{d}(\mathbf{x}, A \setminus \{\mathbf{x}\}) = \frac{1}{|A| - 1} \sum_{\substack{\mathbf{y} \in A \\ \mathbf{y} \neq i}} d(\mathbf{x}, \mathbf{y}), \tag{4.10}$$

where $|A|$ indicates the number of elements in A, and $A \setminus \{\mathbf{v}\}$ implies all elements of A except $\mathbf{v}$.

Also, we must compute the average dissimilarity of element $\mathbf{x}$ to all elements of B:

$$\bar{d}(\mathbf{x}, B) = \frac{1}{|B|} \sum_{\mathbf{y} \in B} d(\mathbf{x}, \mathbf{y}). \tag{4.11}$$

For deciding whether to move this element to cluster B or not, we calculate:

$$D(\mathbf{x}) = \bar{d}(\mathbf{x}, A \setminus \{\mathbf{x}\}) - \bar{d}(\mathbf{x}, B). \tag{4.12}$$

The element with the largest positive D is moved to cluster B. If, however, there is no element that produces $D > 0$, then the division is complete.

For the next steps, we choose the cluster with the largest diameter to split:

$$\text{diam}(C) = \max_{\mathbf{x}, \mathbf{y} \in C} d(\mathbf{x}, \mathbf{y}). \tag{4.13}$$

The process continues until every cluster contains a single element.

To implement DIANA in Julia, we will create sets with all elements and then create a set with these sets:

```
C = Set([M[:,i] for i in :N])
Clusters = Set([C])

for it in 1:N
    # Find largest cluster
    Lmax = -1
    csplit = []    # cluster to be split
    for c in Clusters
        L = diam(c)
        if (L > Lmax)
            Lmax = L
            csplit = c
        end
    end
    # Split cluster if it is not empty
    if (length(csplit) > 1)
        setdiff!(Clusters,[csplit])
        A,B = DIANA_split(csplit)
        push!(Clusters,A)
        push!(Clusters,B)
    end
end
```

We also need a function to split the clusters:

```
# Initial conditions
A = Set(C)
B = Set()

# Find maximum x: d(x,A x)-d(x,B)>0
L = 1
while(L > 0)
    D_max = -1
    v = []
    for x in A
        dA = avg(x,setdiff(A,[x]))
        dB = avg(x,B)
        D = dA - dB
        if (D> 0 && D > D_max)
            D_max = D
            v = x
```

```
        end
    end

    # If Dmax > 0, then A=A\ v, B = B ∪ v
    L = length(v)
    if (L>0)
        setdiff!(A,[v])
        push!(B,v)
    end
end
```

Functions *avg* and *diam* compute the average dissimilarity and the diameter, respectively.

4.4   SPECTRAL CLUSTERING

A graph is an ordered pair $G = (V, E)$, where V is a set of points called vertices (or nodes) and $E \subseteq \{\{x,y\} \mid x,y \in V, x \neq y\}$ is a set of edges (or links) [24]. A graph[c] can be represented by an *adjacency matrix* $\mathbf{A}$ whose elements are the weights between two nodes (typically 1 if they are connected and 0 otherwise). The *Laplacian matrix* for a graph is simply given by $\mathbf{L} = \mathbf{D} - \mathbf{A}$, where $\mathbf{D}$ is the diagonal *degree matrix* whose elements $D_{ii} = \sum_j A_{ij}$ correspond to the number of edges connected to a vertice.

The main idea of spectral clustering algorithms [49] is to construct a graph with the dataset. To accomplish this, one can create an adjacency matrix that establishes a connection between nodes i and j if the latter is among the k-nearest neighbors of the former, and vice-versa. Another possibility is to use a threshold ε to establish a connection between data points. Or, it is also possible to use a fully connected graph and consider weights between nodes given, for example, by a Gaussian function. The latter, in Julia, can be constructed with:

```
for i in 1:(N-1)
    for j in (i+1):N
        d = Euclidean(R[i,1:2],R[j,1:2])
        A[i,j] = exp(-d/(2σ²))
        A[j,i] = A[i,j]
    end
end

for i in 1:N
    D[i,i] = sum(A[i,:])
end
```

---

[c] see Appendix B for an implementation.

The nodes V can be cut into two groups $V^+ \cup V^- = V$ such that the weight of the cut is given by:

$$s(V^+, V^-) = \sum_{\substack{i \in V^+ \\ j \in V^-}} A_{ij} \tag{4.14}$$

We can create a label $y_i \in \{+1, -1\}$ such that $y_i = +1$ if node $i \in V^+$. With this, we can write:

$$s(V^+, V^-) = \frac{1}{4} \sum_{i,j} A_{ij} (y_i - y_j)^2 = \frac{1}{2} \left[\sum_i \left(\sum_j A_{ij} \right) y_i^2 + \sum_{i,j} A_{ij} y_i y_j \right]$$
$$= \frac{1}{2} \mathbf{y}^\top (\mathbf{D} - \mathbf{A}) \mathbf{y}. \tag{4.15}$$

To make calculations simpler, let's use $z_i \in \mathbb{R}$ instead of y_i and construct a map $y_i = \pm 1 \mapsto \text{sign}(z_i) = \pm 1$.

To minimize the cut weight under the constraint that our vector $\mathbf{z}$ is normalized ($1/2 \mathbf{z}^\top \mathbf{D} \mathbf{z} = 1$), we create the Lagrange function:

$$\Lambda = \mathbf{z}^\top (\mathbf{D} - \mathbf{A}) \mathbf{z} - \lambda (\mathbf{z}^\top \mathbf{D} \mathbf{z} - 1). \tag{4.16}$$

Minimizing it with respect to $\mathbf{z}$, we get the eigen-equation:

$$(\mathbf{D} - \mathbf{A}) \mathbf{z} = \lambda \mathbf{D} \mathbf{z}$$
$$(\mathbf{I} - \mathbf{D}^{-1} \mathbf{A}) \mathbf{z} = \lambda \mathbf{z}. \tag{4.17}$$

Note that when $\mathbf{z} = \mathbf{1}$ the product on the left side of the equation gives 0, and we must have the eigenvalue $\lambda = 0$. This, however, corresponds to all elements on the same cluster and this is not what we are after. The eigenvector corresponding to the second eigenvalue (Fiedler[d] vector [50,51]), though, satisfies the balancing condition $\mathbf{z}^\top \mathbf{D} \mathbf{1} = 0$ and defines two clusters whose data can be labeled according to $\text{sign}(\mathbf{z})$. Further clustering can be applied recursively using the same procedure. In Julia:

```
K = I-inv(D)*A
Λ = eigvals(K)
z = sign.(eigvecs(K)[:,2])
```

Figure 4.4 shows the spectral clustering applied to typical non-linearly separable data. The algorithm can successfully cluster the two groups of data, but, unlike DBSCAN is unable to detect noise.

---

[d]Miroslav Fiedler (1926–2015) Czech mathematician.

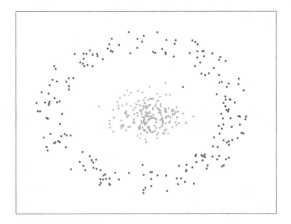

Figure 4.4 A typical non-linearly separable dataset made of two concentric groups of data. A complete graph was formed using a Gaussian weight with $\sigma = 1$

5 Vector Quantization Techniques

In this chapter, we will deal with topological spaces [11] and their quantizations. A topological space can be understood as a set of points $\mathbb{X}$ and a collection of subsets τ (called *topology*) of $\mathbb{X}$ that includes the empty set and $\mathbb{X}$ itself. Moreover, unions and finite intersections of these subsets are also subsets of $\mathbb{X}$. Given this definition, a *manifold* is defined as a topological space whose elements have neighborhoods homeomorphic to a Euclidean space. Examples include circles, surfaces, and spheres.

A k-dimensional submanifold M is defined as the subset $M \subseteq \mathbb{R}^n$ for which there exists a neighborhood $\mathcal{N}(x) \in \mathbb{R}^n$, $\forall x \in M$, and a smooth map $f : U \to \mathbb{R}^n$ for an open set $U \subseteq \mathbb{R}^k$ such that this map f is homeomorphic to $M \cap \mathcal{N}(x)$ and its derivative $\partial_y f$ is injective for every $y \in U$. Or in a narrower sense, a k-dimensional submanifold can be thought of a k-dimensional manifold in the induced topology of the original manifold.

Homeomorphism

Consider a geometric object such as a filled square in the figure below. You can stretch and bend it so that it becomes a disk.

A mapping between two topological spaces that preserves all the topological invariants (for example, the Euler characteristic and the number of holes of a surface, or *genus*) is called a *homeomorphism*. Mathematically, a homeomorphism is defined by a function $f : A \to B$, where A and B are topological spaces, such that f is continuous, bijective, and its inverse is also continuous.

A topological space equipped with the structure of distance between its elements is called a *metric space*.

DOI: 10.1201/9781003350101-5

5.1   SIMPLE VECTOR QUANTIZATION

Vector quantization techniques encode a submanifold M using a finite set of reference vectors $\mathbf{w}_i \in \mathbb{R}^n$ [52]. For an input vector $\mathbf{x} \in M$ in a training set, these algorithms find the node with the closest output $\mathbf{w}_i$:

$$n = \operatorname*{argmin}_i d(\mathbf{x}, \mathbf{w}_i),$$

where d is a metric.

The algorithms then move the winning node closer to the input vector:

$$\begin{aligned}\mathbf{w}_n &\mapsto \mathbf{w}_n + \eta(\mathbf{x} - \mathbf{w}_n) \\ &= \eta\mathbf{x} + (1 - \eta)\mathbf{w}_n,\end{aligned} \tag{5.1}$$

where η is the learning rate.

For the simplest learning vector quantization algorithm in Julia, we create a matrix $\mathbf{X}_{sp}$ whose rows correspond to different dimensions and columns to different points. We also create a matrix $\mathbf{W}$ with the same characteristics corresponding to the weights. The main interaction loop becomes:

```julia
r = rand(1:100)               # Choose vector randomly
x = X_sp[:,r]
n = argmin(d(W,x))            # Find node with closest output
W[:,n]=W[:,n]+η*(x-W[:,n])    # Move winning node toward the input
```

Figure 5.1 shows the result of applying this simple learning vector quantization with 10 points to a sine function after 100, 1000, and 10000 training steps.

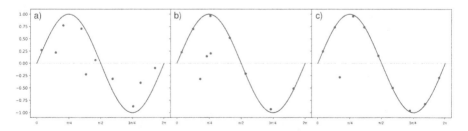

Figure 5.1   Simple learning vector quantization with 10 points applied to a sine function after a) 100, b) 1000, and c) 1000 training steps

5.2   SELF-ORGANIZING MAP

A self-organizing map (SOM) is an idea developed by Kohonen[a] in 1982 [53] where a high-dimensional space maps onto a low-dimensional output space. Hence, it is often known as the *Kohonen map*.

Samples $\mathbf{x}$ of a high-dimensional space are fully connected to a single layer of nodes with weights $\mathbf{w}_i$, $i = 1, \ldots, N$. If those nodes make a bi-dimensional network, then they can be indexed as $\mathbf{w}_{ij}$.

In the training process, each node computes a similarity to the input value through a *discriminant function*. Usually, the Euclidean distance is used:

$$d_{ij} = \|\mathbf{x} - \mathbf{w}_{ij}\|. \tag{5.2}$$

The most similar node to the input (the node with the smallest discriminant function value) is declared the *best matching unit* (BMU). Since nodes compete to be the BMU, this procedure is known as *competitive learning*. Both the BMU and its neighborhood are then moved closer to the input according to:

$$\mathbf{w}_{ij} = \mathbf{w}_{ij} + \eta_t \cdot \mathcal{N}_t^{BMU}(i, j) \cdot (\mathbf{x} - \mathbf{w}_{ij}), \tag{5.3}$$

where $\mathcal{N}_t^{BMU}(i, j)$ is a *topological neighborhood* function typically given by a Gaussian function:

$$\mathcal{N}_t^{BMU}(i, j) = \exp\left(-\frac{(i - i_{BMU})^2 + (j - j_{BMU})^2}{2\sigma_t^2}\right). \tag{5.4}$$

In this last equation, i_{BMU}, j_{BMU} are the coordinates of the BMU in the network, and σ dictates the size of the neighborhood. When the algorithm begins, the neighborhood is large, and most of the points of the output space move, but as the algorithm progresses, we want that the neighborhood decreases to fine-tune the output space. Consequently, this parameter evolves as:

$$\sigma_t = \sigma_0 e^{-t/\tau_\sigma}, \tag{5.5}$$

where σ_0 is set to cover approximately half of the high-dimensional space and τ_σ is a hyperparameter.

Note that the learning rate η_t is also time-dependent. It is desirable for it to progressively decrease over time, allowing for the output space to stabilize. Therefore, we typically use:

$$\eta_t = \eta_0 e^{-t/\tau_\eta}, \tag{5.6}$$

where τ_η is another hyperparameter.

---

[a]Teuvo Kohonen (1934–) Finish engineer, adviser of Erkki Oja.

We finally arrive at an algorithmic view of SOM given by [53]:

Algorithm 7: SOM

---

repeat
$\quad\triangleright$ Choose an input vector $\mathbf{x}$ randomly;
$\quad\triangleright$ Compute each node's discriminant function and find the BMU;
$\quad\triangleright$ Move the BMU and all nodes its neighbors toward the input vector;
$\quad\triangleright$ Update time;
until *desired quantization threshold is achieved*;

Figure 5.2 shows the result of encoding a circle consisting of 200 vectors onto a space consisting of 20 vectors using SOM.

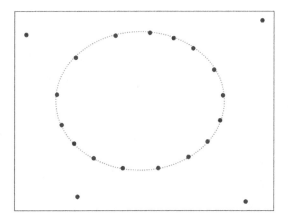

Figure 5.2   The results of encoding a circle consisting of 200 vectors onto a space of 20 vectors. The simulation was performed with 10^5 steps using $\sigma_0 = 0.2$, $\tau_\sigma = 10^5$, $\tau_\eta = 10^5$, and $\eta = 0.1$

5.3   NEURAL GAS

The SOM algorithm has four hyperparameters, and tuning them may be difficult depending on the problem. Furthermore, SOM does not preserve the topological structure of the original space. The neural gas algorithm [54] takes into account the connectivity between nodes through a matrix $\mathbf{C}$. Hence, it produces a lower-dimensionality representation of a sub-manifold preserving its topological structure.

Instead of a winner-takes-all rule, it uses a winner-takes-most rule. For this, the nodes are ordered according to the distance to the input:

$$\mathbf{k} = \underset{i}{\mathrm{argsort}} \|\mathbf{x} - \mathbf{w}_i\|. \tag{5.7}$$

Given an input signal $\mathbf{x}$, the nodes compute excitations $f_i(D_i(\mathbf{x}))$, where $D(\mathbf{w}_i, \mathbf{x})$ is a *distortion* given by $\|\mathbf{x} - \mathbf{w}_i\|$ and the excitation is typically given by $f_i(\mathbf{x}) =$

$\exp(-k_i/\lambda)$, where λ works as a temperature and sets the range of nodes that change their weights in an adaptation step. The weights are then updated as:

$$\Delta\mathbf{w}_i = \eta f(\mathbf{w}_i, \mathbf{x})(\mathbf{x} - \mathbf{w}_i), \tag{5.8}$$

where the learning rate $\eta \in (0, 1]$.

The two nodes in closer proximity to the input are connected by assigning the corresponding position in the $\mathbf{C}$ matrix, while the age of the corresponding edge is reset to zero. All other edges connected to the winner node are aged by one unit. Once an edge reaches a certain threshold age, it is removed.

As the algorithm progresses, the temperature is lowered making the nodes accommodate at definite sites. This typically follows:

$$\eta_t = \eta_0 e^{-t/t_0}, \tag{5.9}$$

where t_0 is a time scale.

Algorithm 8: Neural Gas

---

repeat
> ▷ Choose an input vector $\mathbf{x}$ randomly from the submanifold;
> ▷ Order the distances of the outputs to the input:
>> $\mathbf{k} = \underset{i}{\mathrm{argsort}} \|\mathbf{x} - \mathbf{w}_i\|$
> ▷ Adapt the weights of the topological neighbors of the winning node and of itself:
>> $\Delta\mathbf{w}_i = \eta_t \exp(-k_i/\lambda)(\mathbf{x} - \mathbf{w}_i)$
> ▷ Connect the two nodes closest to the input and set the age of the new edge to zero;
> ▷ Increase the age of the remaining edges connected to k_0;
> ▷ Remove edges older than threshold;

until *desired quantization threshold is achieved*;

5.3.1 GROWING NEURAL GAS

The growing neural gas [55] is an incremental version of the SOM that adjusts its size to the topology that it tries to represent. This is an advantage over SOM, which requires a determination of the number of nodes before running the algorithm.

The structure of this algorithm is very similar to that of SOM. Given an input $\mathbf{x}$, the two nodes s_1 and s_2 closest to the input are found and all edges connected to s_1 are aged. A novel step is inserted by accumulating the error $\|\mathbf{w}_{s_1} - \mathbf{x}\|^2$ into a counter variable $E(s_1)$.

Node s_1 and its topological graph neighbors are moved toward the input:

$$\Delta\mathbf{w}_{s_n} = \eta_n(\mathbf{x} - \mathbf{w}_n). \tag{5.10}$$

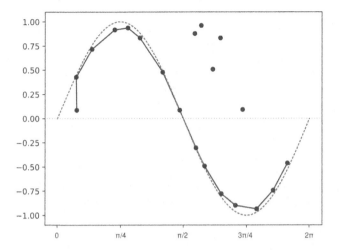

Figure 5.3 Topology constructed by the neural gas algorithm using 20 vectors to reconstruct a sine function

Here η_n, the learning parameter can be made different depending on the order of the node.

An edge $\{s_1, s_2\}$ is created and its age is set to zero. Edges older than a certain threshold are then removed and orphaned nodes are also removed.

Occasionally, a node is inserted. To achieve this, the node s_m with the highest accumulated error and its neighbor with the highest accumulated error s_n are chosen and the edge $\{s_m, s_n\}$ is removed. A node s_h is placed halfway between s_m and s_n:

$$\mathbf{w}_{s_h} = \frac{1}{2}(\mathbf{w}_{s_m} + \mathbf{w}_{s_n}). \tag{5.11}$$

Edges $\{s_h, s_m\}$ and $\{s_h, s_n\}$ are then created. The accumulated errors of nodes s_n and s_m are discounted by a certain amount and $E(s_h)$ is then initialized as $E(s_m)$.

Finally, all errors are discounted and the procedure is repeated until a stopping criterium is reached.

Figure 5.4 shows the results of applying the growing neural gas algorithm to an annulus of inner radius 2 and outer radius 3. Note how the algorithm tries to produce a network that approximates the topology of the submanifold.

5.4   VORONOI AND DELAUNAY DIAGRAMS

Triangulation is the subdivision of a geometric object into simplices (see Sec. 13.1). The Delaunay[b] triangulation is one where no centroid is inside the circumcircle (a circle that passes through all vertices of a polygon) of any simplex, as shown in Fig. 5.5. Upon constructing a circumcircle for the triangle on the left, it promply becomes apparent that a point from the right triangle lies within it, thus violating the Delaunay

---

[b]Boris Delaunay (1890–1980) Russian mathematician, advisee of Georgy Voronoy.

Algorithm 9: Growing Neural Gas

---

repeat
    ▷ Choose an input vector **x** randomly from the submanifold;
    ▷ Find the two nodes s_1 and s_2 closest to the input;
    ▷ Move s_1 and its topological graph neighbors toward **x**;
    ▷ Connect s_1 and s_2 and set the age of the new edge to zero;
    ▷ Age all edges connected to s_1;
    ▷ Add error $\|\mathbf{w}_{s_1} - \mathbf{x}\|^2$ to a counter $E(s_1)$;
    ▷ Remove edges older than a certain threshold. Orphan nodes are also
      removed;
    if *the counter is a multiple of a predetermined value* **then**
        ▷ Insert node s_h halfway between the node with maximum
          accumulated error s_m and its neighbor with largest error s_n;
        ▷ Remove edge $\{s_n, s_m\}$ and include edges $\{s_n, s_h\}$ and $\{s_m, s_h\}$;
        ▷ Discount errors $E(s_n)$ and $E(s_m)$ and then initialize $E(s_h) = E(s_n)$;
    end
    ▷ Reduce all errors by discounting a certain amount;
until *desired quantization threshold is achieved*;

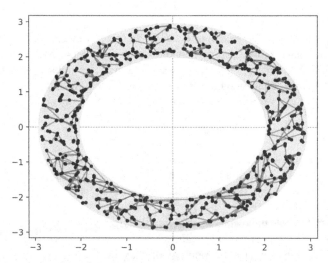

Figure 5.4 Results of 10^4 interactions of the growing neural gas algorithm applied to an annulus of radii 2 and 3

condition. We can fix this problem by flipping the common edge of the triangles as shown in the figure. After this procedure, the two new triangles satisfy the Delaunay condition.

The simplest algorithm [56] used to create Delaunay triangulations is based on a flip operation:

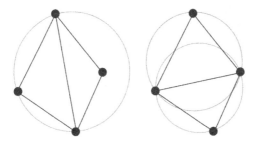

Figure 5.5 Left: two triangles that do not meet the Delaunay conditions, and Right: Flipping the common edge creates two triangles that meet the Delaunay condition

---

Algorithm 10: Flip algorithm

---

input: Any triangulation of a set of points
repeat
  ▷ Flip common edges of non-Delaunay triangles;
until *every triangle is Delaunay*;

The Delaunay triangulation is particularly useful in engineering problems where a mesh is created and finite and boundary element analyses are applied to obtain several properties of a solid.

The dual[c] of the Delaunay triangulation is known as *Voronoi*[d] *tesselation*. A Voronoi region is formed by a collection of points whose distances to a centroid are smaller than the distances to any other centroid. Formally, this is defined as:

$$R_i = \{\mathbf{x} \in \mathbb{X} \mid \forall j \neq i, \, d(\mathbf{x}, \mathbf{C}_i) \leq d(\mathbf{x}, \mathbf{C}_j)\}, \tag{5.12}$$

where $(\mathbb{X}, d)$ compose a metric space and $\mathbf{C}$ is a centroid. If the centroids are known, we can create these regions using the centroids as seeds and radially grow disks until they touch each other. Another possibility is to use the popular Lloyd's algorithm [57]:

Algorithm 11: Lloyd's algorithm

---

input: Data set that can be decomposed in k clusters
  ▷ One can begin with the centroids of the clusters as seeds;
repeat
  ▷ Compute k Voronoi cells Ω_i using the seeds;
  ▷ Compute the centroids of the constructed Voronoi cells;
  ▷ Move the seeds toward these centroids;
until *no more movement is observed*;

---

[c]The dual of a planar graph is another graph that has a vertex for each face of the original graph.
[d]Georgy Voronoy (1868–1908) Ukranian mathematician, advisee of Andrey Markov and adviser of Boris Delaunay and Wacław Sierpiński.

Section III

Supervised Learning

6 Regression Models

Here we begin to discuss supervised learning. This corresponds to a set of algorithms that learn how to establish connections between inputs and outputs, often leveraging labeled data. Once this correspondence is made, the algorithms can generalize and make predictions on new data. Supervised learning finds applications in image classification and speech recognition, for example.

In the context of regression models, which is the topic of this chapter, the significance of visualizing the data prior to any analysis is crucial to the actual calculations. Consider, for example, the four plots in Fig. 6.1 known as the *Anscombe*[a] *quartet* [58].

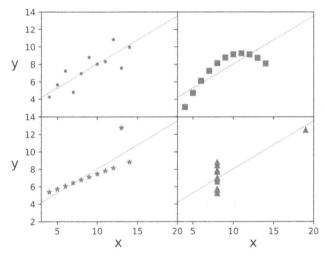

Figure 6.1 The four datasets have the same average values, the same variances of the x and y components, the same correlation between x and y, and the same linear regression coefficients. Nonetheless, they are visually different

Although the datasets in all cases have the same average values, the same variances of the x and y components, the same correlation between them, and the same linear regression coefficients, they are visually different. The plots on the bottom display the contribution of outliers in a regression, whereas the upper right plot shows how a non-linear relationship could mistakenly be fitted with a linear model.

Consider now the plot in Fig. 6.2. Although the whole collection of data can display a negative trend, we find positive trends within clusters of data. This

---

[a]Francis John Anscombe (1918–2001) English statistician.

contradiction, known as *Simpson's*[b] *paradox* [59] also illustrates the need for data visualization before fitting the data.

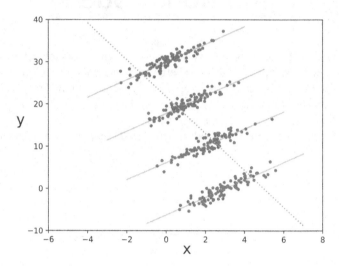

Figure 6.2 Simpson's paradox shows different trends for the whole data set and within data clusters

In this chapter, we will study three major types of regression: linear, logistic, and non-linear. But first, we must take a look at how the complexity of the adopted model may lead to problems in fitting the data.

6.1 BIAS-VARIANCE TRADE-OFF

We have a true function $f(x)$, a set of data given by $y = f(x) + \eta$, where η is a zero-centered random variable with variance σ^2, and an estimator $\hat{f}(x)$ that approximates $f(x)$.

The bias of an estimator is defined as the difference between its expected value and the value of the true function:

$$\text{Bias}_{\hat{f}} = \langle \hat{f}(x) \rangle - f(x) = \langle \hat{f}(x) - f(x) \rangle. \tag{6.1}$$

Its variance, on the other hand, is given by:

$$\text{Var}_{\hat{f}} = \left\langle \left(\hat{f}(x) - \langle \hat{f}(x) \rangle \right)^2 \right\rangle. \tag{6.2}$$

Finally, the mean square error (MSE) related to this estimator is given by:

$$\text{MSE} = \left\langle \left(f(x) - \hat{f}(x) \right)^2 \right\rangle = \text{Bias}_{\hat{f}}^2 + \text{Var}_{\hat{f}} + \sigma^2. \tag{6.3}$$

---

[b]Edward Hugh Simpson (1922–2019) British statistician.

This is known as the bias-variance decomposition [60] of the mean squared error because, for a fixed MSE, a reduction in bias is counter with an increase in variance and vice-versa.

Underfitting happens when the model is incapable of establishing a relationship between input and output variables accurately. This often leads to a high loss on both the training and validation losses (see more details in Sec. 8.2.1). This can be the result of a model being too simple and unable to account for the complexity of the data. Therefore, this can be overcome by training the model for longer periods or adding more complexity to the model. Typically one identifies underfitting when there is high bias and low variance in the model.

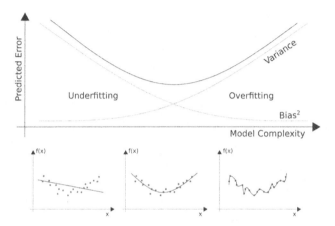

Figure 6.3   a) Bias, variance, and the mean squared error as functions of the model complexity, b) underfitting corresponding to high bias and low variance, c) good fitting that finds an adequate compromise between bias and variance, and d) overfitting corresponding to low bias and high variance

Overfitting, on the other hand, happens when the model is overly complex and captures details of the training data that do not necessarily represent the main trend. Consequently, the model is unable to generalize well to new data and may lead to low training losses but high validation losses. This can be identified when there is low bias but high variance. Usually, overfitting can be reduced by collecting more training data or using regularization, which we will discuss in Sec. 6.4.

6.2   SIMPLE LINEAR REGRESSION

Imagine a machine whose input is controlled by some knob such that each measurement is independent of each other. We test N input values x_i, $i = 1,\dots,N$ and record their corresponding output values y_i. We would like to build a model that gives an estimated output given a certain input. The first model we can think of consists of a linear relationship between input and output. Thus, we construct an estimator $\hat{y}_i = ax_i + b$, where a and b are parameters to be determined. To facilitate our calculations, we can group all these data in vectors $\mathbf{x}$, $\mathbf{y}$, and $\hat{\mathbf{y}}$ such that we can write:

$$\hat{\mathbf{y}} = a\mathbf{x} + b\mathbf{e}, \tag{6.4}$$

where $\mathbf{e}$ is a vector where all elements are 1.

Under this description, the mean square error produced by our model is given by:

$$L = \frac{1}{N}(\mathbf{y} - \hat{\mathbf{y}})^2 = \frac{1}{N}(a\mathbf{x} + b\mathbf{e} - \mathbf{y})^2. \tag{6.5}$$

We would like our model to produce the smallest error possible. Therefore, we minimize it with respect to the parameters of the model to obtain, in matrix notation [3,61]:

$$\begin{bmatrix} \frac{1}{N}\mathbf{x}\cdot\mathbf{x} & \frac{1}{N}\mathbf{x}\cdot\mathbf{e} \\ \frac{1}{N}\mathbf{e}\cdot\mathbf{x} & \frac{1}{N}\mathbf{e}\cdot\mathbf{e} \end{bmatrix} \begin{bmatrix} a \\ b \end{bmatrix} = \begin{bmatrix} \langle\mathbf{x}^2\rangle & \langle\mathbf{x}\rangle \\ \langle\mathbf{x}\rangle & 1 \end{bmatrix} \begin{bmatrix} a \\ b \end{bmatrix} = \begin{bmatrix} \frac{1}{N}\mathbf{x}\cdot\mathbf{y} \\ \frac{1}{N}\mathbf{e}\cdot\mathbf{y} \end{bmatrix}. \tag{6.6}$$

Inverting this matrix, we get:

$$\begin{bmatrix} a \\ b \end{bmatrix} = \frac{1}{\sigma_x^2} \begin{bmatrix} \langle\mathbf{xy}\rangle - \langle\mathbf{x}\rangle\langle\mathbf{y}\rangle \\ \langle\mathbf{x}^2\rangle\langle\mathbf{y}\rangle - \langle\mathbf{x}\rangle\langle\mathbf{xy}\rangle \end{bmatrix}, \tag{6.7}$$

where σ_x^2 is the variance of the data in $\mathbf{x}$ and we are assuming that it is constant along the whole time series (*homoscedasticity*).

Parameter a can be readily recognized as:

$$a = \frac{\sigma_{xy}^2}{\sigma_x^2}, \tag{6.8}$$

where σ_{xy}^2 is the covariance between $\mathbf{x}$ and $\mathbf{y}$.

Parameter b can be easily obtained from the second line of the matrix in Eq. 6.6:

$$b = \langle\mathbf{y}\rangle - a\langle\mathbf{x}\rangle. \tag{6.9}$$

Using Julia, a simple regression can be computed with:

```
using Statistics

a = cov(x,y)
b = mean(y)-a*mean(x)
```

An example of linear regression is shown in Fig. 6.4.

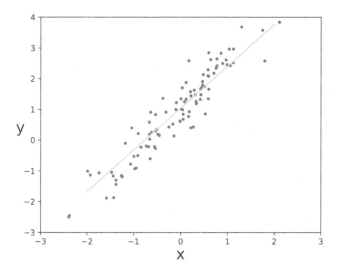

Figure 6.4 Simple linear regression. Solid disks are training data, whereas the gray line is the inferred linear model

6.2.1 GOODNESS OF FIT

The quality of fit can be quantified by the correlation coefficient between $\mathbf{y}$ and $\hat{\mathbf{y}}$ [61]:

$$r_{y\hat{y}} = \frac{\sigma^2_{y\hat{y}}}{\sqrt{\sigma^2_y \sigma^2_{\hat{y}}}}. \tag{6.10}$$

Since $\hat{\mathbf{y}}$ is an estimator of $\mathbf{y}$, we can write $\mathbf{y} = \hat{\mathbf{y}} + \eta\mathbf{e}$, where η is a zero-centered random variable.

If we square this correlation coefficient, we get:

$$r^2_{y\hat{y}} = \frac{\left(\sigma^2_{\hat{y}+\eta,\hat{y}}\right)^2}{\sigma^2_y \sigma^2_{\hat{y}}} = \frac{\left(\sigma^2_{\hat{y},\hat{y}} + \sigma^2_{\eta,\hat{y}}\right)^2}{\sigma^2_y \sigma^2_{\hat{y}}}. \tag{6.11}$$

But assuming that the random variable η is not correlated with the signal that we are estimating (*exogeneity*), we get:

$$r^2_{y\hat{y}} = \frac{\sigma^2_{\hat{y}}}{\sigma^2_y}. \tag{6.12}$$

This is also known as the *coefficient of determination*. This can be written as:

$$r^2 = \frac{[\sum_i (y_i - \langle y\rangle)(\hat{y}_i - \langle y\rangle)]^2}{\sum_i (y_i - \langle y\rangle)^2 \sum_j (\hat{y}_j - \langle y\rangle)^2} = \frac{\sum_i (\hat{y}_i - \langle y\rangle)^2}{\sum_i (y_i - \langle y\rangle)^2}. \tag{6.13}$$

The term in the numerator is known as *explained variance*, whereas the term in the denominator is known as *total variance*. Furthermore, the *unexplained variance* is the difference between the total and explained variances such that we can write:

$$r^2 = 1 - \frac{\text{unexplained variance}}{\text{total variance}}. \qquad (6.14)$$

Hence, the ratio quantifies the variability in y that is not accounted for by the regression. Consequently, we aim for values of r^2 close to 1, which imply a good fit.

In Julia, we can simply do:

```
r2 = cor(y,yh)^2
```

The coefficient of determination for the data in Fig. 6.4, for example, is ~90%.

6.3 MULTIVARIATE LINEAR REGRESSION

Suppose now that, instead of a single knob, we have k different knobs that control the input of our machine, and the knobs control an input data set $\mathbf{x}_i$, $i = 1, \ldots, k$. Our model becomes [61]:

$$\hat{\mathbf{y}} = a_0 \mathbf{e} + a_1 \mathbf{x}_1 + a_2 \mathbf{x}_2 + \ldots + a_k \mathbf{x}_k. \qquad (6.15)$$

Again, in order to make calculations simpler, we can group all in a single matrix:

$$\hat{\mathbf{y}} = \begin{bmatrix} \mathbf{e} & \mathbf{x}_1 & \cdots & \mathbf{x}_k \end{bmatrix} \begin{bmatrix} a_0 \\ a_1 \\ \vdots \\ a_k \end{bmatrix} = \mathbf{Xa}, \qquad (6.16)$$

where $\mathbf{X}$ is known as the *design matrix*.

Again, we write a loss:

$$L = \frac{1}{N}(\hat{\mathbf{y}} - \mathbf{y})^2 = \frac{1}{N}(\mathbf{Xa} - \mathbf{y})^2 \qquad (6.17)$$

and minimize it with respect to $\mathbf{a}$:

$$\frac{\partial L}{\partial \mathbf{a}} = \frac{1}{N}\mathbf{X}^\top(\mathbf{Xa} - \mathbf{y}) = 0 \qquad (6.18)$$

Solving for $\mathbf{a}$, we obtain:

$$\mathbf{X}^\top \mathbf{Xa} = \mathbf{X}^\top \mathbf{y}$$
$$\mathbf{a} = \left(\mathbf{X}^\top \mathbf{X}\right)^{-1} \mathbf{X}^\top \mathbf{y} \qquad (6.19)$$
$$= \mathbf{X}^+ \mathbf{y},$$

where $X^+ = (X^\top X)^{-1} X^\top$ is known as the Moore[c]-Penrose[d] pseudoinverse of X [62,63]. Moreover, the term $X^\top X$ is a *Gram*[e] *matrix*[f], and $X^\top y$ is known as a *moment matrix*.

Once we have obtained $\mathbf{a}$, we can estimate the output of our machine given some input X:

$$\hat{y} = Xa = XX^+ y = Py, \tag{6.20}$$

where P is known as the *projection matrix* that projects y onto $\hat{y}$. Therefore, the residuals can be computed as $y - \hat{y} = (I - P)y = My$, where M is known as the *annihilator matrix*.

An example of multivariate regression with two predictors x and y is shown in Fig. 6.5. Coefficient r^2, in this case, was $\sim 98\%$. In Julia, this kind of regression can be calculated with:

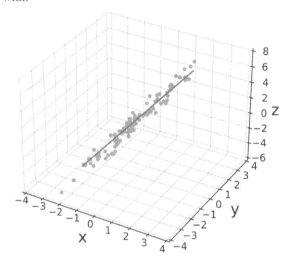

Figure 6.5 Multivariate linear regression with two predictors x and y. Solid disks are training data, whereas the solid line is the predicted model

```
lbd = 0.1
Gram = X'*X + I*lbd
MP = inv(Gram)*X'
a = MP*y
```

[c]Eliakim Hastings Moore (1862–1932) North American mathematician, adviser of George Birkhoff, among others.
[d]Roger Penrose (1931–Present) British mathematician. Penrose won several awards including the Nobel Prize in Physics in 2020.
[e]Jørgen Pedersen Gram (1850–1916) Danish mathematician.
[f]A symmetric and positive semidefinite Hermitian matrix given by the inner products of a set of vectors.

6.3.1 BAYESIAN APPROACH

Another way to obtain the same result is using Bayes inference [64]. For this, we use, as a prior:

$$P(\mathbf{a}) = \exp\left\{-\frac{1}{2}\mathbf{a}^\top(\sigma_0^{-2}\mathbf{I})\mathbf{a}\right\}. \tag{6.21}$$

The likelihood of finding $\mathbf{y}$ given parameters $\mathbf{a}$ is given by:

$$P(\hat{\mathbf{y}}|\mathbf{a}) = \exp\left\{-\frac{1}{2}(\hat{\mathbf{y}} - \mathbf{Xa})^\top(\sigma^{-2}\mathbf{I})(\hat{\mathbf{y}} - \mathbf{Xa})\right\}. \tag{6.22}$$

Using Bayes rule, the posterior is given by:

$$\begin{aligned} P(\mathbf{a}|\hat{\mathbf{y}}) \propto P(\mathbf{a})P(\hat{\mathbf{y}}|\mathbf{a}) &= \exp\left\{(\mathbf{a} - \mu)^\top\Sigma^{-1}(\mathbf{a} - \mu)\right\} \\ &= \exp\left\{-\frac{1}{2}\left[\hat{\mathbf{y}}^\top\sigma^{-2}\hat{\mathbf{y}} - \hat{\mathbf{y}}^\top\sigma^{-2}\mathbf{Xa} - \mathbf{a}^\top\mathbf{X}^\top\sigma^{-2}\hat{\mathbf{y}} + \right. \right. \\ &\qquad\qquad \left.\left. +\mathbf{a}^\top\mathbf{X}^\top\sigma^{-2}\mathbf{Xa} + \mathbf{a}^\top\sigma_0^{-2}\mathbf{a}\right]\right\} \end{aligned} \tag{6.23}$$

Inspecting this last expression, we find:

$$\Sigma^{-1} = \sigma^{-2}\mathbf{X}^T\mathbf{X} + \sigma_0^{-2}\mathbf{I} \tag{6.24}$$

and

$$\Sigma^{-1}\mu = \mathbf{X}^\top\sigma^{-2}\hat{\mathbf{y}}$$
$$\therefore \mu = \left[\mathbf{X}^T\mathbf{X} + \left(\frac{\sigma_0}{\sigma}\right)^{-2}\mathbf{I}\right]^{-1}\mathbf{X}^\top\hat{\mathbf{y}}. \tag{6.25}$$

If we compare this expected value for the parameters $\mathbf{a}$ with that in Eq. 6.19, we see that there is an extra term $\lambda = (\sigma_0/\sigma)^{-2}$. Consider a situation where the product $\mathbf{X}^\top\mathbf{X}$ is composed of small values. This damping term modifies the pseudoinverse by making the Gram matrix more well-conditioned and, consequently, easier to invert. Small values of λ make the estimation of the parameters more sensitive to the values of the moment matrix, which corresponds to more complex models. On the other hand, large values of λ stabilize the inversion, constraining the solutions to a smaller set of values, and, consequently, making the estimation of the parameters less sensitive to the values of the moment matrix. This corresponds to less complex models and is known as *regularization*, which will be discussed next.

6.4 REGULARIZATION

Many times, the problems we try to solve can be ill-conditioned. For example, it may happen that one input can be predicted by other inputs (collinearity) or we might have many more variables than observations. These problems may make the Gram matrix $\mathbf{X}^\top\mathbf{X}$ nearly singular and sensitive to small changes in $\mathbf{X}$. We can reduce this variance by increasing the bias of the model. For this, we create a Lagrange function for the loss [61]:

$$\Lambda = (\mathbf{Xa} - \mathbf{y})^2 + (\Gamma \mathbf{a})^2, \tag{6.26}$$

where Γ is known as the *Tikhonov*[g] *matrix* [65].

Minimizing it with respect to **a**:

$$\frac{\partial \Lambda}{\partial \mathbf{a}} = \mathbf{X}^\top (\mathbf{Xa} - \mathbf{y}) + \Gamma^\top \Gamma \mathbf{a} = 0$$

$$\mathbf{a} = \left(\mathbf{X}^\top \mathbf{X} + \Gamma^\top \Gamma\right)^{-1} \mathbf{X}^\top \mathbf{y}. \tag{6.27}$$

The Tikhonov matrix can be chosen to be a diagonal matrix of the form $\Gamma = \sqrt{\lambda} \mathbf{I}$, $\lambda \geq 0$ that shifts the diagonal elements of the Gram matrix, making it better conditioned. This *ridge regression* gives preference to solutions with small norms. In this case, the Lagrange function in Eq. 6.26 can be simplified as:

$$\Lambda = (\mathbf{Xa} - \mathbf{y})^2 + \lambda \left(\sum_{i=0}^{k} a_i^2 - c \right), \tag{6.28}$$

where λ is known as the *ridge parameter* and c is a target value.

Another regularization possibility is to penalize the absolute value of each element of **a** instead of its squared values:

$$\Lambda = (\mathbf{Xa} - \mathbf{y})^2 + \lambda \left(\sum_{i=0}^{k} |a_i| - c \right). \tag{6.29}$$

This is known as the *least absolute shrinkage and selection operator* (LASSO) [66, 67].

From a geometric standpoint, we can consider a convex function, represented by $(\mathbf{Xa} - \mathbf{y})^2$, with two distinct constraints. In the case of LASSO, these constraints create a hypercube, while for ridge regression, they form a k-sphere. When dealing with the same convex function, it is more probable for the hypercube to intersect with the function along one of the axes, rather than the k-sphere achieving a similar alignment. This relationship is illustrated in Fig. 6.6, which visualizes the concept in two dimensions. Consequently, LASSO tends to facilitate dimensionality reduction.

6.4.1 LEAST ANGLE REGRESSION

Unfortunately, the LASSO regularization is not differentiable and, consequently, an analytical expression for **a** cannot be obtained. Nevertheless, it is possible to compute LASSO solutions using some algorithms such as the *least-angle regression* (LARS) [68] that we will discuss next. First, however, let's take a look at the *stagewise* algorithm.

---

[g] Andrey Nikolayevich Tikhonov (1906–1993) Russian mathematician.

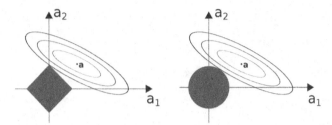

Figure 6.6   The shaded regions correspond to the restrictions for LASSO (left) and ridge (right). The ellipses correspond to $(\mathbf{Xa} - \mathbf{y})^2$

For the stagewise algorithm, we begin with a null model $\mathbf{a} = 0$ and make the residuals $\mathbf{r} = \hat{\mathbf{y}}$. Next, we compute the correlations $\rho_i = \langle r, \mathbf{x}_i \rangle$ between each predictor and the residuals and choose the predictor with the largest correlation coefficient (and consequently, one that results in the lowest residual sum of squares). We then update $a_i \leftarrow a_i + \delta_i$, where $\delta_i = \eta \cdot \text{sign}(\rho_i)$ and η is a positive and small number. Finally, we update the residuals $\mathbf{r} \leftarrow \mathbf{r} - \delta_i \mathbf{x}_i$ and go back to computing correlations until some stopping criterium has been achieved.

In summary, the algorithm reads:

Algorithm 12: LARS

$\triangleright$ Set $\mathbf{a} = \mathbf{0}$ and $\mathbf{r} = \hat{\mathbf{y}}$;
repeat
    $\triangleright$ Find $i = \underset{j}{\text{argmax}}(\mathbf{x}_j^\top \mathbf{r})$;
    $\triangleright$ Update $a_i \leftarrow a_i + \delta_i$, where $\delta_i = \eta \cdot \text{sign}(\rho_i)$;
    $\triangleright$ Update $\mathbf{r} \leftarrow \mathbf{r} - \delta_i x_i$;
until *desired performance is obtained*;

LARS is an improvement of this algorithm. The initial conditions and the first two steps are the same. The increase in a_i in step 3, though, continues until some other predictor $\mathbf{x}_k$ has the same correlation. The update happens for both predictors and is not in the direction of $\mathbf{x}_i$ but in an equiangular direction between $\mathbf{x}_i$ and $\mathbf{x}_k$. The process continues until all predictors have entered the computation or $\mathbf{x}_s^\top \mathbf{r} = 0$, $\forall s$. If a coefficient crosses zero, though, it is removed from the active set, the direction is recomputed and the procedure continues.

In Julia, the main iteration loop of LARS begins with:

```
ρ = [sum(r), x1·r, x2·r]
i = argmax(abs.(ρ))

if (∼(i in lst))
    push!(lst,i)
end
```

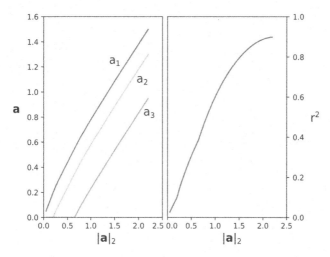

Figure 6.7 Evolution of parameters **a** (left) and r^2 (right) as a function of the norm of **a**

Then, the equiangular direction is found:

```
d = 0
for j in lst
    d = d + η*sign(ρ[j])
end
d = d / length(lst)
```

Finally, the coefficients are updated:

```
for j in lst
    a[j] = a[j] + d
    r = r - d*X[:,j]
end
```

6.5 NON-LINEAR REGRESSION

Consider a function $f : \mathbb{R} \to \mathbb{R}$ with inputs x_i, $i = 1,\ldots,N$ and parameters **w**. Let's now take the second-order multivariate Taylor's[h] expansion for this function

---

[h]Brook Taylor (1685–1731) English mathematician.

around $\mathbf{w}_k$:

$$f(x_i; \mathbf{w}_{k+1}) \approx f(x_i; \mathbf{w}_k) + \frac{\partial f(x_i; \mathbf{w}_k)}{\partial \mathbf{w}_k}(\mathbf{w}_{k+1} - \mathbf{w}_k). \tag{6.30}$$

This can be recast in matrix notation as:

$$\mathbf{f}_{k+1} \approx \mathbf{f}_k + \mathbf{J}(\mathbf{w}_{k+1} - \mathbf{w}_k), \tag{6.31}$$

where $\mathbf{J}$ is the Jacobian matrix for function f.

As in the case of the standard linear regression, let's create an objective function corresponding to the square error for real data y_i:

$$S = |\mathbf{y} - \mathbf{f}_k - \mathbf{J}(\mathbf{w}_{k+1} - \mathbf{w}_k)|^2. \tag{6.32}$$

Minimizing this loss for $\mathbf{w}_{k+1}$, we get:

$$\frac{\partial S}{\partial \mathbf{w}_{k+1}} = -2\mathbf{J}^\top \left(\mathbf{y} - \mathbf{f}_k - \mathbf{J}(\mathbf{w}_{k+1} - \mathbf{w}_k)\right) = 0 \tag{6.33}$$
$$\mathbf{J}^\top \mathbf{J}(\mathbf{w}_{k+1} - \mathbf{w}_k) = \mathbf{J}^\top(\mathbf{y} - \mathbf{f}_k).$$

These are known as *normal equations* and can be solved for $\mathbf{w}_{k+1}$:

$$\mathbf{w}_{k+1} \approx \mathbf{w}_k + \left(\mathbf{J}^\top \mathbf{J}\right)^{-1} \mathbf{J}^\top (\mathbf{y} - \mathbf{f}_k). \tag{6.34}$$

Note that, to compute the updated weights, it is necessary to compute the pseudoinverse (see Sec. 6.3) of the Jacobian[i] matrix. This method for updating the weights is known as Gauss[j]-Newton[k] method [4].

It may happen, though that the matrix $\mathbf{J}^\top \mathbf{J}$ is not invertible. This can be fixed by introducing a regularization:

$$\mathbf{w}_{k+1} \approx \mathbf{w}_k + \left(\mathbf{J}^\top \mathbf{J} + \lambda \mathbf{I}\right)^{-1} \mathbf{J}^\top (\mathbf{y} - \mathbf{f}_k), \tag{6.35}$$

where $\lambda > 0$ is a damping parameter that interpolates between the Gauss-Newton method and the direction of the steepest descent.

Gradient Descent

Consider a function $f : \mathbb{R}^n \to \mathbb{R}$. Its infinitesimal change for any infinitesimal displacement $d\mathbf{r} \in \mathbb{R}^n$ is given by:

$$df = \nabla f \cdot d\mathbf{r}.$$

Therefore, the maximal change of function f occurs when the infinitesimal

---

[i]Carl Gustav Jacob Jacobi (1804–1851) German mathematician, adviser of Otto Hesse, among others.
[j]Johann Carl Friedrich Gauss (1777–1855) German mathematician, adviser of Bernhard Riemann among others.
[k]Isaac Newton (1642–1726) English natural philosopher, adviser of Roger Cotes, among others.

Gradient Descent (Continued)

displacement $d\mathbf{r}$ is in the direction of the gradient ∇f. Therefore, it is possible to move to higher values of the function by choosing points in the direction of its gradient. Conversely, one can choose progressively smaller values of f moving away from its gradient. If the function is differentiable in the neighborhood of some point θ, then this approach can be described by:

$$\theta_{n+1} = \theta_n - \eta \nabla_\theta f(\theta_n), \tag{6.36}$$

where $\eta \in \mathbb{R}_+$, known as the *learning rate*, specifies how fast a point moves away from the gradient. This is known as the *gradient descent* algorithm or the *steepest descent* algorithm.

This algorithm can be improved by scaling each element of the diagonal maintaining the curvature of the gradient:

$$\mathbf{w}_{k+1} \approx \mathbf{w}_k + \left(\mathbf{J}^\top \mathbf{J} + \lambda \operatorname{diag}(\mathbf{J}^\top \mathbf{J}) \right)^{-1} \mathbf{J}^\top (\mathbf{y} - \mathbf{f}_k). \tag{6.37}$$

This is known as the Levenberg[l]-Marquardt[m] algorithm [69, 70].

Example 6.5.1. Levenberg-Marquardt

As an example, let's consider a model given by the function $f(x) = a \cdot \sin(bx)$. In Julia:

```julia
f = w[1]*sin.(x*w[2])
```

The derivatives of this function for both parameters are:

$$\frac{\partial f(x)}{\partial a} = \sin(bx)$$
$$\frac{\partial f(x)}{\partial b} = ax \cdot \cos(bx) \tag{6.38}$$

In Julia, we write:

```julia
J1 = sin.(w[2]*x)
J2 = w[1]*x.*cos.(w[2]*x)
J = hcat(J1,J2)
```

---

[l]Kenneth Levenberg (1919–1973) American statistician.
[m]Donald Marquardt (1929–1997) American statistician.

The Levenberg-Marquardt loop is obtained with:

w = w + (inv(J'*J)*J')*(y-f)

The result of this procedure is shown in Fig. 6.8. ∎

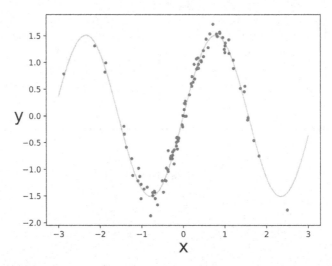

Figure 6.8 A training dataset (solid disks) and a predicted non-linear model (solid curve)

6.6 GAUSSIAN PROCESS REGRESSION

In section 6.3.1 we used Bayes' theorem to do a linear regression. Here, instead of having a model, we will fit functions to data using a *Gaussian process*, which is a collection of random variables that have a joint Gaussian distribution [71, 72]. A Gaussian process is completely specified by a mean function μ and a positive definite covariance function. The covariance function is given by a kernel function that describes how far two points are. Usually, a *radial basis function* (RBF) is used:

$$k(\mathbf{x}, \mathbf{y}) = \sigma_0^2 \exp\left\{ \frac{-\|\mathbf{x} - \mathbf{y}\|^2}{2\lambda^2} \right\}, \tag{6.39}$$

where σ_0^2 is the *signal variance* and λ is the *length scale*.

This can be implemented in Julia as:

```julia
function kernel(x,y)
    σ0 = 1.0
    λ = 0.6

    L2(x) = sum(x.^2)

    Nx = length(x)
    Ny = length(y)

    K = Array-Float64"(undef,Nx,Ny)

    for i in 1:Nx
        for j in 1:Ny
            K[i,j] = σ0*exp(-L2(x[i]-y[j])/2λ^2)
        end
    end

    return K
end
```

Consider a set of input points $\mathbf{x} = \begin{bmatrix} x_1 & x_2 & \dots & x_N \end{bmatrix}$ and a function f that gives a collection of outputs $\mathbf{f} = \begin{bmatrix} f(x_1), f(x_2), \dots, f(x_N) \end{bmatrix}$. We are interested in, given a set of other points $\mathbf{x}_*$, estimating the function $\mathbf{f}_*$ at those points. To find it, we need to compute the conditional distribution $P(\mathbf{f}_*|\mathbf{f}, \mathbf{x}, \mathbf{x}_*)$, but right now we only have the joint distribution of $\mathbf{f}$ and $\mathbf{f}_*$:

$$P(\mathbf{f}, \mathbf{f}_*|\mathbf{x}, \mathbf{x}_*) \sim \mathcal{N} \left(\begin{bmatrix} \mu(\mathbf{x}) \\ \mu(\mathbf{x}_*) \end{bmatrix}, \begin{bmatrix} k(\mathbf{x}, \mathbf{x}) + \sigma_e^2 \mathbf{I} & k(\mathbf{x}, \mathbf{x}_*) \\ k(\mathbf{x}_*, \mathbf{x}) & k(\mathbf{x}_*, \mathbf{x}_*) \end{bmatrix} \right), \tag{6.40}$$

where σ_e^2 is the noise level of the observations. The posterior is given by conditioning the joint distribution [34, 73], which is a normal distribution, with:

$$\begin{aligned} \mu &= k(\mathbf{x}_*, \mathbf{x}) \left[k(\mathbf{x}, \mathbf{x}) + \sigma_e^2 \mathbf{I} \right]^{-1} \mathbf{f} \\ \Sigma &= k(\mathbf{x}_*, \mathbf{x}_*) - k(\mathbf{x}_*, \mathbf{x}) \left[k(\mathbf{x}, \mathbf{x}) + \sigma_e^2 \mathbf{I} \right]^{-1} k(\mathbf{x}, \mathbf{x}_*). \end{aligned} \tag{6.41}$$

Once we have the mean and the covariance matrix, we can generate samples. However, for this, we need the standard deviation matrix. Fortunately, the covariance matrix is positive-definite. Therefore, we can apply Cholesky[n] decomposition,

---

[n] André-Louis Cholesky (1875–1918) French mathematician.

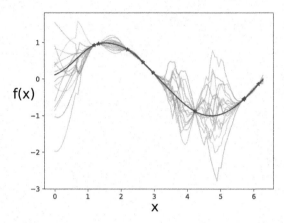

Figure 6.9 Observed points (black stars), the computed mean (solid black line), and different samples

which states that $\Sigma = LL^{\top}$, where L is a lower triangular matrix [3,4]. In Julia:

```
L = cholesky(Symmetric(Σ),check=false).L
g = μ + L*randn(M)
```

A set of samples calculated for observations drawn from a sine curve are shown in Fig. 6.9.

7 Classification

Classification of input data to predefined categories or classes is another main goal of supervised learning. These algorithms typically learn a decision boundary in the form of adjustments of a model. One of the most used algorithms is based on using a logistic function to probabilistically classify data. Classification trees work by recursively splitting the data into a tree-like structure, where each node represents a feature, and each end leaf is assigned to a different class label. After discussing trees, we will study support vector machines, which are algorithms that try to find a hyperplane that separates different classes of data. Finally, the chapter ends with the naïve Bayes algorithm, which is a simple yet useful algorithm based on the Bayes' theorem.

7.1 LOGISTIC REGRESSION

The idea of logistic regression is to find a relationship between a set of predictors (features or independent variables) and a random variable that produces the probability of some label. Consider, for example, a Bernoulli[a] trial characterized by a random variable Y such that the probability of success is given by p and the probability of failure is given by $1 - p$ [74]. Also, suppose that we know that the expected value of Y is p_0. According to the principle of maximum entropy, we can find the distribution that best represents our knowledge about this trial by maximizing the Lagrange function:

$$\Lambda = -\sum_{i=0}^{1} P(Y = i) \log\left(P(Y = i)\right) + \lambda\left(\langle Y \rangle - p_0\right)$$
$$= -(1 - p)\log(1 - p) - p\log(p) + \lambda(p - p_0), \tag{7.1}$$

where λ is a Langrange multiplier. The maximization of this function gives:

$$\log\left(\frac{p}{1 - p}\right) = \lambda \to \frac{p}{1 - p} = e^{\lambda}. \tag{7.2}$$

This is the same as saying that we know the odds ratio[b] (or the strength of association between the probabilities of success and failure), given by the exponential of λ.

After a little algebra, we get:

$$p = \frac{1}{1 + e^{-\lambda}}, \tag{7.3}$$

---

[a]Jacob Bernoulli (1655–1654) advised by Gottfried Wilhelm Leibniz, adviser of Nicolaus I and Johann Bernoulli.
[b]The log-odds is also known as the *logit* function (short for "**logi**stic **unit**").

that is the *logistic function*[c]: $\mathbb{R} \to (0,1)$ [75], which gives the probability of success given a Lagrange multiplier λ.

Given a set of explanatory variables $\{x_i\}$, $i = 1,\ldots,N$, it is convenient to write the Lagrange multiplier λ as a linear combination of these variables:

$$p(\mathbf{x}) = \frac{1}{1 + e^{-\mathbf{w} \cdot \mathbf{x}}}, \tag{7.4}$$

where $\mathbf{w}$ is a vector of weights and $\mathbf{x} = \begin{bmatrix} 1 & x_1 & \ldots & x_N \end{bmatrix}^\top$. With this, we can estimate the result of a Bernoulli trial given this set of explanatory variables. This behavior is known as a *binary classifier* [76].

Now we need to estimate $\mathbf{w}$ to use this classifier. The probability mass function for the Bernoulli distribution over possible outcomes k is given by $p^k(1-p)^{1-k}$, $k \in \{0,1\}$. For a set of data points $\{\mathbf{x}_i, y_i\}$, $i = 1,\ldots,N$, the likelihood and the log-likelihood functions are given by:

$$\mathcal{L} = \prod_{i=1}^{N} p(\mathbf{x}_i)^{y_i} \left(1 - p(\mathbf{x}_i)\right)^{1-y_i},$$
$$\log(\mathcal{L}) = \sum_{i=1}^{N} y_i \log(p(\mathbf{x}_i)) + (1 - y_i) \log(1 - p(\mathbf{x}_i)). \tag{7.5}$$

Unfortunately, minimizing the log-likelihood does not provide a closed-form solution. Nonetheless, we can use the Newton-Raphson method to find a solution.

Newton-Raphson Method

Let's begin with second-order multivariate Taylor's expansion for a function $f : \mathbb{R}^p \to \mathbb{R}$:

$$f(\mathbf{w}_{k+1}) \approx f(\mathbf{w}_k) + \frac{\partial f(\mathbf{w}_k)}{\partial \mathbf{w}_k}(\mathbf{w}_{k+1} - \mathbf{w}_k) + \\ + \frac{1}{2}(\mathbf{w}_{k+1} - \mathbf{w}_k)^\top \mathbf{H}(\mathbf{w}_{k+1} - \mathbf{w}_k), \tag{7.6}$$

where $\mathbf{J} = \frac{\partial f}{\partial \mathbf{w}_k}$ is the Jacobian matrix and $\mathbf{H}$ is the Hessian matrix. Taking its derivative for $\mathbf{w}_{k+1}$, one gets:

$$\frac{\partial f(\mathbf{w}_{k+1})}{\partial \mathbf{w}_{k+1}} \approx \frac{\partial f(\mathbf{w}_k)}{\partial \mathbf{w}_k^\top} + \frac{1}{2}\left(\mathbf{H}(\mathbf{w}_{k+1} - \mathbf{w}_k) + \mathbf{H}^\top(\mathbf{w}_{k+1} - \mathbf{w}_k)\right). \tag{7.7}$$

Given that the Hessian matrix is symmetric:

$$\frac{\partial f(\mathbf{w}_{k+1})}{\partial \mathbf{w}_{k+1}} \approx \frac{\partial f(\mathbf{w}_k)}{\partial \mathbf{w}_k^\top} + \mathbf{H}(\mathbf{w}_{k+1} - \mathbf{w}_k). \tag{7.8}$$

---

[c]Developed by Pierre François Verhulst (1804–1849) Belgian mathematician, advised by Lambert Adolphe Jacques Quetelet.

> ### Newton-Raphson Method (Continued)
>
> The second-order Taylor expansion is minimized when:
>
> $$\mathbf{w}_{k+1} \approx \mathbf{w}_k - \mathbf{H}^{-1} \frac{\partial f(\mathbf{w}_k)}{\partial \mathbf{w}_k^\top}. \tag{7.9}$$
>
> This is known as the Newton-Raphson[a] method for optimization [3, 4].
>
> ---
> [a]See Section 3.4 for a first-order version.

To apply the Newton-Raphson method, we need to compute the gradient and the Hessian[d] of the log-likelihood function. The gradient is given by:

$$\frac{\partial \log(\mathscr{L})}{\partial \mathbf{w}} = \sum_{i=1}^{N} \left(\frac{y_i}{p(\mathbf{x}_i)} - \frac{1 - y_i}{1 - p(\mathbf{x}_i)} \right) \frac{\partial p(\mathbf{x}_i)}{\partial \mathbf{w}}. \tag{7.10}$$

To move further, we need:

$$\frac{\partial p(\mathbf{x}_i)}{\partial \mathbf{w}} = \frac{\partial}{\partial w_j} \left(1 + e^{-\mathbf{w} \cdot \mathbf{x}_i} \right)^{-1} = p(\mathbf{x}_i)(1 - p(\mathbf{x}_i)) \mathbf{x}_i^\top. \tag{7.11}$$

Consequently, the derivative of the log-likelihood function becomes:

$$\frac{\partial \log(\mathscr{L})}{\partial \mathbf{w}^\top} = \mathbf{X}^\top (\mathbf{y} - \mathbf{p}(\mathbf{x})), \tag{7.12}$$

where

$$\mathbf{X} = \begin{bmatrix} 1 & (\mathbf{x}_1)_1 & (\mathbf{x}_1)_2 & \dots & (\mathbf{x}_1)_p \\ 1 & (\mathbf{x}_2)_1 & (\mathbf{x}_2)_2 & \dots & (\mathbf{x}_2)_p \\ \vdots & \vdots & \vdots & \ddots & \vdots \\ 1 & (\mathbf{x}_N)_1 & (\mathbf{x}_N)_2 & \dots & (\mathbf{x}_N)_p \end{bmatrix}. \tag{7.13}$$

The Hessian function is given by:

$$\frac{\partial^2 \log(\mathscr{L})}{\partial \mathbf{w} \partial \mathbf{w}^\top} = - \sum_{i=1}^{N} \mathbf{x}_i \frac{\partial}{\partial \mathbf{w}} p(\mathbf{x}_i) = -\mathbf{X}^\top \mathbf{S} \mathbf{X}, \tag{7.14}$$

where $\mathbf{S}$ is a diagonal matrix with elements $p(\mathbf{x}_i)(1 - p(\mathbf{x}_i))$ and we have used the result of Eq. 7.11 on the second line.

Applying these results to Newton-Raphson, we get:

$$\mathbf{w}_{k+1} \approx \mathbf{w}_k + \left(\mathbf{X}^\top \mathbf{S} \mathbf{X} \right)^{-1} \mathbf{X}^\top (\mathbf{y} - \mathbf{p}). \tag{7.15}$$

---
[d]Ludwig Otto Hesse (1811-1874) German mathematician, advised by Carl Gustav Jacob Jacobi, and adviser of Carl Neumann and Gustav Kirchhoff, among others.

This can also be written as:

$$\mathbf{w}_{k+1} \approx \left(\mathbf{X}^\top \mathbf{S} \mathbf{X}\right)^{-1} \mathbf{X}^\top \mathbf{S} \left(\mathbf{X}\mathbf{w}_k + \mathbf{S}^{-1}(\mathbf{y} - \mathbf{p})\right)$$
$$\approx \left(\mathbf{X}^\top \mathbf{S} \mathbf{X}\right)^{-1} \mathbf{X}^\top \mathbf{S} \mathbf{z}, \tag{7.16}$$

where $\mathbf{z} = \mathbf{X}\mathbf{w}_k + \mathbf{S}^{-1}(\mathbf{y} - \mathbf{p})$ is known as the *adjusted response*. Note that this is the solution to the weighted least squares problem:

$$\mathbf{w}_{k+1} \leftarrow \underset{\mathbf{w}_k}{\mathrm{argmin}} \, (\mathbf{z} - \mathbf{X}\mathbf{w}_k)^\top \mathbf{S}(\mathbf{z} - \mathbf{X}\mathbf{w}_k), \tag{7.17}$$

where the weights are given by matrix $\mathbf{S}$. Hence, this is known as *iteratively reweighted least squares* (IRLS) [77].

In Julia, the main iteration loop of the IRLS algorithm for a two-dimensional model can be implemented as shown below. Here, $\mathbf{X} = \begin{bmatrix} 1 & \mathbf{x}_1 & \mathbf{x}_2 \end{bmatrix}$ and $\mathbf{y}$ contain the input and output training data, whereas $\mathbf{w}$ contains the weights:

```
p = [1.0/(1.0 + exp(-w·r)) for i in eachrow(X)]
S = diagm(p)
w = w + inv(X'*S*X)*X'*(y-p)
```

A good procedure to test the goodness of fit for the logistic regression is calculating the *deviance*. This is defined as:

$$D = -2\log\left(\frac{\mathscr{L}(\mathbf{x}|\mathbf{w})}{\mathscr{L}(\mathbf{x}|\mathbf{w}_o)}\right), \tag{7.18}$$

where $\mathbf{w}_o$ indicates the fitted parameters for the *saturated model* (a model with a perfect fit). The likelihood for the saturated model is identically equal to 1. Therefore, the deviance becomes:

$$D = -2\log\left(\mathscr{L}(\mathbf{x}|\mathbf{w})\right). \tag{7.19}$$

Model deviance and null deviance can also be computed. The latter is computed with only the first value w_1. The difference between both gives an idea of how well the predictors improve the model:

$$\Delta D = -2\log\left(\frac{\mathscr{L}(\mathbf{x}|w_1)}{\mathscr{L}(\mathbf{x}|\mathbf{w}_o)}\right) + 2\log\left(\frac{\mathscr{L}(\mathbf{x}|\mathbf{w})}{\mathscr{L}(\mathbf{x}|\mathbf{w}_o)}\right)$$
$$= -2\log\left(\frac{\mathscr{L}(\mathbf{x}|w_1)}{\mathscr{L}(\mathbf{x}|\mathbf{w})}\right). \tag{7.20}$$

This is χ^2 distributed with $p - 1$ degrees of freedom [78]. The same procedure can be used to test the contribution of each predictor.

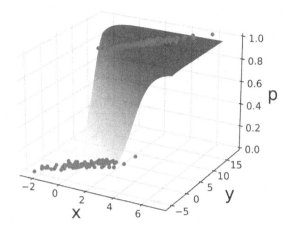

Figure 7.1   Training dataset (solid disks) and the corresponding logistic regression (surface)

The quality of the fit can be estimated in Julia with:

```
L = 1
for (qi,yi) in zip(p,y)
    L = L*(pi∧yi)*((1-pi)∧(1-yi))
end

ΔD = -2*log(L)
```

7.2   CLASSIFICATION AND REGRESSION TREES

Decision trees [79] are a powerful and simple prediction approach based on undirected connected acyclic graphs where decisions are taken at each node of such graph. It tends to be simple to understand as it mirrors human decision-making. To make a decision tree, data is commonly structured in a table. Each column consists of some attribute, except the last one which is a class with labels that we want to infer.

For example, consider an investor that wants to decide between buying a stock. To make a decision, he considers i) if the price of the stock is low, high, or fair, and ii) if the potential growth of that stock is also low, high, or fair. Then, the investor organizes his investing strategy in a table as shown in Tab. 7.1. Here, the attributes are price and potential, whereas the only class is to invest or not. Consequently, its two possible labels are yes and no. The data presented in the table is a set of training examples that we will denote by $T : \{(\mathbf{x}, y)\}$, where $\mathbf{x}$ is a vector of attribute values and y is the label of the target class.

Table 7.1

Example of Structured Data for Constructing a Decision Tree

Price	Potential	Invest
low	low	no
low	high	yes
high	low	no
high	high	no
fair	high	yes
fair	fair	yes

To construct a tree, we begin by computing the Shannon entropy [13] of the classes without considering the attributes:

$$H_0 = - \sum_{i=\text{label}} p_i \log_2(p_i)$$

$$= -\frac{3}{6} \log_2 \left(\frac{3}{6}\right) - \frac{3}{6} \log_2 \left(\frac{3}{6}\right) = 1, \tag{7.21}$$

where we used the fact that three possibilities lead to investing and also three possibilities lead to not investing.

In Julia, we can create a function that computes the Shannon entropy. The first thing inside this function is to find the number of labels present in the target class. Here, v is the input, n is initialized with zero and counts the number of labels, and l is an array of labels:

```
for vi in v
    if ~(vi in l)
        push!(l,vi)
        n = n + 1
    end
end
```

Next, we find the frequencies of each label and, for this, p is an array of frequencies:

```
for i in 1:length(v)
    for j in 1:n
        if (v[i]==l[j])
            p[j]=p[j]+1
        end
    end
end
```

Finally, we compute the entropy:

```
p = p/s
H = -sum(p.*log2.(p))

return H
```

The information that we gain considering an attribute a is given as:

$$
\begin{aligned}
G(T,a) &= H(T) - H(T|a) \\
&= H(T) - \langle H(S_a) \rangle_a \\
&= H(T) - \sum_{i \in \text{vals}(a)} p_i(v) H(S_a(i)),
\end{aligned}
\tag{7.22}
$$

where

$$
S_a(i) = \{\mathbf{x} \in T | x_a = i\}.
\tag{7.23}
$$

We can create another function in Julia to compute the information gain. This will have inputs T, the training set, and p, which corresponds to the attribute. First, we find the number of labels for attribute p, which is the same algorithm presented above. Again, n is the number of labels, and l is an array of labels. Next, we must compute the frequencies and relative entropies:

Let's apply this to our example. If we consider the attribute "price," we can write the following table:

The entropy for attribute value "low" was computed as:

$$
S_{\text{low}} = -\frac{1}{2} \log_2 \left(\frac{1}{2}\right) - \frac{1}{2} \log_2 \left(\frac{1}{2}\right) = 1.
\tag{7.24}
$$

Since the conditional probabilities $p(yes|high) = p(no|fair) = 0$, the respective entropies are zero. The information gain is consequently $G(T, \text{price}) = 1 - 1/3 = 2/3 = 0.67$.

For the attribute "potential," we create Tab. 7.3. The same argument that we used for the previous attribute works here, and we end up with an information gain $G(T, \text{potential}) = 1 - (1/2) \cdot 0.92 = 0.54$.

Table 7.2

Table of Conditional Events Considering Attribute "price" for Constructing a Decision Tree

	Yes	No	Total	p_i	S
Low	1	1	2	2/6	1
High	0	2	2	2/6	0
Fair	2	0	2	2/6	0
	3	3	6	1	

Table 7.3

Table of Conditional Events Considering Attribute "potential" for Constructing a Decision Tree

	Yes	No	Total	p_i	S
Low	0	2	2	2/6	0
High	2	1	3	3/6	0.92
Fair	1	0	1	1/6	0
	3	3	6	1	

Since we gain more information with attribute the "price," we use it as a *root node*. Also, given that the entropy is zero for attribute values "high" and "fair," they are *leaf nodes*. Attribute value "low" needs to be further split. We see from the original data that when "price" is "low" we have "potential" with a value "low" not leading to an investment recommendation and a value "high" leading to an investment recommendation. This completes our tree in Fig. 7.2, and summarizes the *iterative dichotomiser 3* (ID3) algorithm[e] [80]:

Algorithm 13: ID3

---

repeat
 ▷ For every feature $x \in \mathbf{x} \in T$, find the entropy;
 ▷ Split T into subsets using the attribute for which the gain of
  information is the highest;
 ▷ Create a node with this attribute;
 ▷ Apply the same strategy recursively on the subsets with the remaining
  attributes;
until *all elements in the subset belong to the same class, or all attributes were selected, or there are no examples in the subset*;
 ▷ Turn node into a leaf;

---

[e]Developed by John Ross Quinlan, Australian computer scientist.

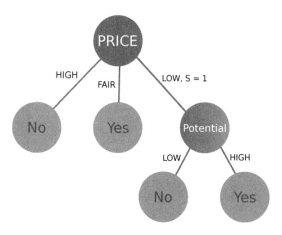

Figure 7.2 Decision tree built with the training set T shown in Tab. 7.1

7.2.1 BOOSTING

Trees can become overly complex and lead to overfitting. This can be overcome by either pruning the decision tree or inferring from an ensemble. In the latter strategy, training sets are created by sampling from the original dataset T with replacement. This process is known as *bootstrap aggregating* or *bagging* [81], for shorter.

A collection of different and independent decision trees are then created for each training set. When estimating, a voting procedure is performed to obtain an output. If we are dealing with regression trees, then voting consists of taking the average over all trees, whereas if we are dealing with classification trees, then voting consists of choosing the majority label.

One problem with bagging is that, if using the same training algorithm, the trees might end up being highly correlated. The *random forest* [82,83] algorithm mitigates this effect by instead of considering all features in a split, considering only a smaller random subset, typically around the square root of the total number of features.

It is possible to do boosting adaptively with an algorithm called *AdaBoost* [84]. In this algorithm, we have a set of features x_i, $i = 1, \ldots, N$ with a corresponding class $y_i \in [-1, 1]$, and the idea is to generate a set of weak classifiers $k_i(x_i) \in [-1, 1]$. After n iterations, the booster classifier is given by:

$$C_n(x_i) = \sum_{j=1}^{n} \alpha_j k_j(x_i), \tag{7.25}$$

where α is a collection of weights that we will call *performance*. Now we add a new classifier:

$$C_{n+1}(x_i) = C_n(x_1) + \alpha_{n+1} k_{n+1}(x_i). \tag{7.26}$$

We use the sum of the exponential loss as the total error:

$$\mathcal{E} = \sum_{i=1}^{N} e^{-y_i C_{n+1}(x_i)} = \sum_{i=1}^{N} e^{-y_i C_n(x_i)} e^{-y_i \alpha_{n+1} k_{n+1}(x_i)} = \sum_{i=1} w_{i,n} e^{-y_i \alpha_{n+1} k_{n+1}(x_i)}, \quad (7.27)$$

where

$$w_{i,n} = \begin{cases} 1 & n=1 \\ e^{-y_i C_n(x_i)} & n>1. \end{cases} \quad (7.28)$$

Let's now split this sum between those that were misclassified ($y_i k_{n+1}(x_i) = -1$) and those that were correctly classified ($y_i k_{n+1}(x_i) = 1$) by the weak classifier:

$$\mathcal{E} = \sum_{y_i=k_i} w_{i,n} e^{-\alpha_{n+1}} + \sum_{y_i \neq k_i} w_{i,n} e^{\alpha_{n+1}}. \quad (7.29)$$

Minimizing this error with respect to α_{n+1}, we obtain:

$$\frac{\partial \mathcal{E}}{\partial \alpha_{n+1}} = -\sum_{y_i=k_i} w_{i,n} e^{-\alpha_{n+1}} + \sum_{y_i \neq k_i} w_{i,n} e^{\alpha_{n+1}} = 0. \quad (7.30)$$

This leads to:

$$\alpha_{n+1} = \frac{1}{2} \log \left(\frac{\sum_{y_i=k_i} w_{i,n}}{\sum_{y_i \neq k_i} w_{i,n}} \right). \quad (7.31)$$

Since the error is normalized to add up to 1, we can compute the error rate:

$$\varepsilon_n = \frac{\sum_{y_i \neq k_i} w_{i,n}}{\sum_i w_{i,n}} \quad (7.32)$$

and end up with:

$$\alpha_{n+1} = \frac{1}{2} \log \left(\frac{1-\varepsilon_n}{\varepsilon_n} \right), \quad (7.33)$$

which is half of the logit function (see page 103).

Let's see how this algorithm works with an example. Consider the data in Tab. 7.4. The weight for each row is initialized with $1/N$.

First, in Julia, we must create a data structure for a stump:

```julia
mutable struct stumpStructure
    attribute :: Int
    out :: Dict{String,Int}
end
```

This weak classifier uses only *attribute* to classify data, and its output is given by the dictionary *out*. The stump is trained (*fitBinaryStructure*) so that it outputs +1 if the label of the respective attribute leads to more labels +1 of the output class, and

Table 7.4
Hypothetical Dataset for AdaBoost

Feature 1	Feature 2	Feature 3	Output	Weight
a	c	e	yes (+1)	1/5
b	d	e	yes (+1)	1/5
a	c	e	no (-1)	1/5
b	c	e	no (-1)	1/5
a	d	g	yes (+1)	1/5

-1 otherwise. The output of the stump is given by the function *stump*. The classifier for AdaBoost is an implementation of Eq. 7.26:

```
function Classifier(c1,c2,c3,Ensemble,alphas)
    C = 0
    for (s,α) in zip(Ensemble,alphas)
        C = C + stump(s,c1,c2,c3)*α
    end

    return sign(C)
end
```

We now create a function that adds a new stump to the ensemble. First, the function creates a new stump and fits it according to the experimental data:

```
function Add(Ensemble,alphas,weights,p,T)
    s = stumpStructure(p,Dict())
    fitBinaryStump!(T,s)
```

Next, the function finds the data that are misclassified by the stump so that the error rate can be computed. Also, it stores the classification output in the list ξ_list to be used later:

```
misclassified = 0
total = 0
ξ_list = []
for (xi,w) in zip(eachrow(T),weights)
    k = stump(s,xi[1],xi[2],xi[3])

    y = xi[4]
    if (k ≠ y)
        misclassified += w
        push!(ξ_list,1)
    else
        push!(ξ_list,−1)
    end
    total += w
end

ε = misclassified/total
α = 0.5*log((1-ε)/ε)
```

Now the weights are updated and the function returns the new ensemble, alphas, and the new weights. The weight for each row is updated to:

$$w_{n+1} = \frac{w_n}{Z}e^{\xi\alpha}, \tag{7.34}$$

where $\xi = 1$ if the rows misclassified the data and $\xi = -1$ otherwise. Z is a normalizing constant such that the updated weights sum up to 1:

```
nw = []
Z = 0
for (w,ξ) in zip(weights,ξ_list)
    el = w*exp(ξ*α)
    Z = Z + el
    push!(nw,el)
end
nw = nw/Z

return vcat(Ensemble,s), vcat(alphas,α), nw
end
```

When running the snippet, we see that stumps that use attributes 1 and 3 are very weak classifiers. Their information gains are only 0.02 and 0.17, respectively. Therefore, an ensemble containing only these two stumps can only achieve a performance of 60 %. When a stump that uses attribute 2 for classification is added, the performance of the ensemble goes to 80 %.

In summary, the AdaBoost algorithm reads:

Algorithm 14: AdaBoost

▷ Initialize weights to $1/n$;
for t *in 1:N* **do**
   ▷ Fit a weak learner (stump) k_t to the data;
   ▷ Compute the classification error rate ε and $\alpha_t = 1/2 \log\left(\frac{1-\varepsilon}{\varepsilon}\right)$;
   ▷ Add it to the ensemble $C_t(\mathbf{r}) = C_{t-1}(\mathbf{r}) + \alpha_t k_t(\mathbf{r})$;
   ▷ Update weights: $w_{i,t+1} = w_{i,t} e^{\xi_i \alpha_t}/Z$;
end

7.3 SUPPORT VECTOR MACHINES

A hyperplane [85], illustrated in Fig. 7.3, can be defined by a point $\mathbf{r}_0$ and a normal vector $\mathbf{n}$, such that, for any point $\mathbf{r}$ that lies on this hyperplane, we have:

$$\mathbf{n}\cdot(\mathbf{r} - \mathbf{r}_0) = 0. \tag{7.35}$$

Since $\mathbf{r}_0$ is a constant, we can rewrite this equation as:

$$\mathbf{n}\cdot\mathbf{r} + b = 0, \tag{7.36}$$

where $b = -\mathbf{n}\cdot\mathbf{r}_0$.

The idea of this technique [86] is to use this hyperplane as a classifier with a function:

$$f(\mathbf{x}) = \text{sgn}(\mathbf{n}\cdot\mathbf{x} + b). \tag{7.37}$$

The argument must be positive for points i whose labels d_i are $+1$, and negative for points whose labels are $d_i = -1$.

Now, the distance of a point $\mathbf{x}$ to this hyperplane is given by:

$$r = \frac{\mathbf{n}\cdot(\mathbf{x}-\mathbf{r}_0)}{\|\mathbf{n}\|} = \frac{\mathbf{n}\cdot\mathbf{x} + b}{\|\mathbf{n}\|}. \tag{7.38}$$

To make calculations simple, let's impose that:

$$\mathbf{n}\cdot\mathbf{x} + b \begin{cases} \geq +1 & \text{if } d_i = +1 \\ \leq -1 & \text{if } d_i = -1. \end{cases} \tag{7.39}$$

These boundaries refer to the two hyperplanes shown in Fig. 7.3. The separation between these hyperplanes, as defined by Eq. 7.38, is denoted by:

$$s = \frac{2}{\|\mathbf{n}\|}. \tag{7.40}$$

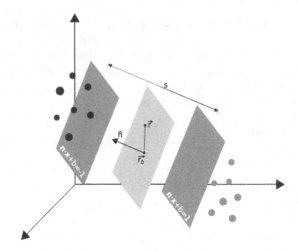

Figure 7.3   A hyperplane characterized by a normal vector **n**, a point $\mathbf{r}_0$, and any other point **r**. The image also contains two other hyperplanes described by Eq. 7.39 that separate two sets of data, one with dark dots and another with light dots

Vectors that lie on these two hyperplanes are known as *support vectors*.

We would like, then, to find these two hyperplanes that maximize the separation s. This, however, is equivalent to minimizing $\|\mathbf{n}\|$. Furthermore, we have two restrictions imposed by Eq. 7.39 that can be simplified to the feasibility condition:

$$d_i\left(\mathbf{n}\cdot\mathbf{x}_i+b\right)\geq 1. \tag{7.41}$$

We now have a cost function and a constraint, both conditions that lead to the use of Lagrange multipliers. For this, we create a Lagrange function:

$$\mathscr{L}=\frac{1}{2}\|\mathbf{n}\|^2-\sum_i\lambda_i\left[d_i\left(\mathbf{n}\cdot\mathbf{x}_i+b\right)-1\right], \tag{7.42}$$

where $\lambda_i > 0$ are Lagrange multipliers [87]. We particularly used a quadratic term because it creates a paraboloid surface with a well-defined single global minimum.

Minimizing it with respect to **n** we get the stationarity condition:

$$\mathbf{n}=\sum_i\lambda_i d_i\mathbf{x}_i. \tag{7.43}$$

And minimizing it with respect to b, we get:

$$\sum_i\lambda_i d_i=0. \tag{7.44}$$

Instead of using these equations to solve this *quadratic programming* problem [88], let's invert the problem by plugging the equations back into the Lagrange function:

$$\mathscr{L} = \frac{1}{2} \sum_i \lambda_i d_i \mathbf{n} \cdot \mathbf{x}_i - \sum_i \lambda_i \left[d_i \left(\mathbf{n} \cdot \mathbf{x}_i + b \right) - 1 \right]$$

$$= -\frac{1}{2} \sum_i \lambda_i d_i \mathbf{n} \cdot \mathbf{x}_i - \sum_i \lambda_i \qquad (7.45)$$

$$= -\frac{1}{2} \sum_{i,j} \lambda_i \lambda_j d_i d_j \mathbf{x}_i \cdot \mathbf{x}_j - \sum_i \lambda_i.$$

This is known as the *dual problem*, whereas the original Lagrange function is known as the *primal problem* [87,88]. The dual problem is particularly interesting because it is a function only of the training set pairs $(\mathbf{x}_i, d_i)$ and includes the dot product, which can be used to include kernels.

7.3.1   SOFT MARGINS

When the problem is not separable, the multipliers may diverge, but we typically want bounded solutions. Therefore, we can relax the linear separability restriction by inserting *slack variables* $(\xi_i > 0)$ that are related to the degree of violation of the constraint. Now, Eq. 7.41 becomes:

$$d_i \left(\mathbf{n} \cdot \mathbf{x}_i + b \right) \geq 1 - \xi_i, \ \xi_i > 0. \qquad (7.46)$$

The amount $\sum_i \xi_i$ can be understood as a penalty for errors, and we would like it to be as small as possible. Thus, the primal problem becomes:

$$\mathscr{L} = \frac{1}{2} \|\mathbf{n}\|^2 + C \sum_i \xi_i - \sum_i \lambda_i \left[d_i \left(\mathbf{n} \cdot \mathbf{x}_i + b \right) - 1 + \xi_i \right] - \sum_i \mu_i \xi_i, \qquad (7.47)$$

where μ_i and C are new multipliers. The former was introduced to guarantee that $\xi_i > 0$, and the latter was introduced to include the penalty for having a *soft margin*.

The stationarity condition gives:

$$\frac{\partial \mathscr{L}}{\partial \mathbf{n}} = \mathbf{n} - \sum_i \lambda_i d_i \mathbf{x}_i = 0$$

$$\frac{\partial \mathscr{L}}{\partial b} = \sum_i \lambda_i d_i = 0$$

$$\frac{\partial \mathscr{L}}{\partial \xi_i} = C - \lambda_i + \mu_i = 0 \qquad (7.48)$$

$$\frac{\partial \mathscr{L}}{\partial \lambda_i} = - \left[d_i \left(\mathbf{n} \cdot \mathbf{x}_i + b \right) - 1 - \xi_i \right] = 0.$$

From the third line, we see that $\lambda_i = C - \mu_i$, but $\mu_i \geq 0$. Therefore, C is an upper bound for λ_i, and the dual problem is now given by:

$$\begin{cases} \mathscr{L} = \dfrac{1}{2}\sum_{i,j}\lambda_i\lambda_j d_i d_j K(\mathbf{x}_i,\mathbf{x}_j) - \sum_i \lambda_i \\ \text{s.t. } \sum_i \lambda_i d_i = 0 \\ 0 \le \lambda_i \le C. \end{cases} \qquad (7.49)$$

There are many techniques to solve this optimization problem. One of the most popular is the *sequential minimization optimization* [89] which is discussed in Appendix C.

7.4   NAIVE BAYES

The last classifier we will discuss in this chapter is the Naïve Bayes classifier [90]. Suppose you are deciding whether you should invest in an asset, and gathered some historical information described in Tab. 7.5.

Table 7.5

Hypothetical Feature Matrix and Response Vector Used in a Nave Bayes Classifier

Price	Volume	Bought
10.3	100	no
9.5	50	yes
11.5	90	no
5.5	30	yes
7.5	120	yes
12.2	40	no
7.1	80	yes
10.5	65	no

The data in the first three columns of the table correspond to the *feature matrix*, whereas the data in the last column correspond to the *response vector*. We will use the notation $\mathbf{X} = \begin{bmatrix} \mathbf{x}_1 & \mathbf{x}_2 & \cdots & \mathbf{x}_N \end{bmatrix}$ for the former, and $\mathbf{y}$ for the latter.

Our objective is to construct a classifier that can determine, based on a set of features, whether we should buy an asset. For this, we will use Bayes' rule (see Sec. 2.1.2.1.):

$$P(y|\mathbf{x}) = \frac{P(\mathbf{x}|y)P(y)}{P(\mathbf{x})}. \qquad (7.50)$$

To proceed, we will use the naïve hypothesis that the evidence is composed of independent data and their contributions are equally important. Therefore, we may write:

Table 7.6

Parameters μ and σ Calculated for the Data from Tab. 7.5

Action	μ_{price}	σ^2_{price}	μ_{volume}	σ^2_{volume}
Buy	7.40	2.71	70.00	1533.33
Not	11.12	0.79	73.75	722.92

$$P(y|\mathbf{x}) = P(y|x_1, x_2, \ldots, x_N)$$
$$= \frac{[P(x_1|y)P(x_2|y)\ldots P(x_N|y)]\,P(y)}{P(x_1)P(x_2)\ldots P(x_N)} \tag{7.51}$$
$$\propto P(y) \prod_{i=1}^{N} P(x_i|y).$$

In our example, the prior is:

$$\text{prior}(y = \text{yes}) = \text{prior}(y = \text{no}) = \frac{4}{8}. \tag{7.52}$$

The posteriors are given by:

$$\text{posterior}(\text{buy}) = \text{prior}(\text{buy})P(\text{Price}|\text{buy})P(\text{Volume}|\text{buy}),$$
$$\text{posterior}(\text{not}) = \text{prior}(\text{not})P(\text{Price}|\text{not})P(\text{Volume}|\text{not}). \tag{7.53}$$

To compute the posteriors, we need to make an assumption about the distribution from which the data is collected. This assumption is referred to as an *event model*, and, typically, the Gaussian and the Bernoulli distributions are used. In the case of the Gaussian distribution, the conditional probability is written as:

$$P(x|y) = \frac{1}{\sqrt{2\pi\sigma_y^2}} \exp\left\{ -\frac{(x - \mu_y)^2}{2\sigma_y^2} \right\}, \tag{7.54}$$

where μ_y and σ_y^2 are the mean and the unbiased variance of the features associated with class y. In our case, these parameters are given in Tab. 7.6.

This probability is called *maximum posterior* (MAP), and the classifier is given by the value of y that maximizes it:

$$y = \arg\max_y P(y) \prod_{i=1}^{n} P(x_i|y). \tag{7.55}$$

For our example, we obtain the map illustrated in Fig. 7.4.

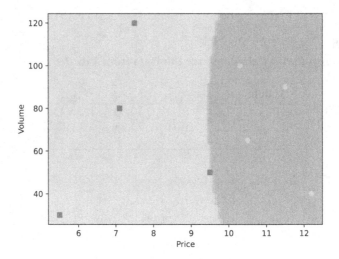

Figure 7.4 Decision map obtained using the naïve Bayes classifier with the data from Tabs. 7.5 and 7.6. Squares are observed data corresponding to buying the asset, whereas disks are observed data corresponding to not buying the asset. The clear region corresponds to the classifier recommending the purchase of the asset, and the dark region corresponds to the classifier not recommending the purchase

In Julia, we can write:

```
function Classifier(price,volume,p)
    μᵖʸ, μᵖⁿ, μᵥʸ, μᵥⁿ, σᵖ²ʸ, σᵖ²ⁿ, σᵥ²ʸ, σᵥ²ⁿ = p

    P_buy = Gaussian(price,μᵖʸ,σᵖ²ʸ)*Gaussian(volume,μᵥʸ,σᵥ²ʸ)
    P_not = Gaussian(price,μᵖⁿ,σᵖ²ⁿ)*Gaussian(volume,μᵥⁿ,σᵥ²ⁿ)

    if P_buy > P_not
        return 1
    else
        return 0
    end
end
```

Despite its simplicity, the naïve Bayes classifier performs incredibly well and is easy to implement for big data. On the other hand, because of the simplistic assumptions, it can be inaccurate sometimes, and other models may outperform it if abundant data is available.

Section IV

Neuronal-Inspired Learning

8 Feedforward Networks

Consider a continuous function shown in Fig. 8.1. This function can be approximated as the sum of several rectangular functions and each rectangular function can be built from the difference between two offset step functions. With this approach, then, it should be possible to create an approximation $\hat{f}(x)$ of any continuous function $f(x)$ such that $\|\hat{f}(x) - f(x)\| < \varepsilon$ for any input x and arbitrarily small ε. It is also possible to improve this approximation by using progressively narrower rectangles.

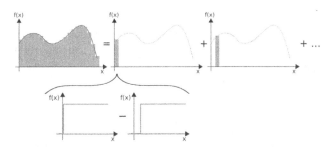

Figure 8.1 The decomposition of a continuous function into a set of rectangular functions. Each rectangular function can be obtained from the subtraction of two step functions

We have particularly chosen a rectangular function, but this approximation could be made with many other similar functions. Instead of the subtraction of two step functions, we could use the difference between two other non-linear functions as long as there is an offset between them. This offset, though, could be generalized as an affine map of the input. Moreover, the Kolmogorov[a]-Arnold[b] representation theorem [91,92] allows us to extend this approach to multivariate continuous functions. Thus, we arrive at a building block:

$$g(\mathbf{x}) = \sigma(\mathbf{W}\mathbf{x} + \mathbf{b}), \tag{8.1}$$

where $\mathbf{W}$ is a linear map, known as the *weight matrix*, $\mathbf{b}$ is a constant vector known as *bias*, and σ is a non-linear function known as the *activation function*. By combining these building blocks, it should be possible to approximate any continuous function[c].

---

[a]Andrey Nikolaevich Kolmogorov (1903–1987) Russian mathematician, winner of many prizes, adviser of Vladimir Arnold, among others.

[b]Vladimir Igorevich Arnold (1937–2010) Russian mathematician winner of many prizes, advised by of Andrey Kolmogorov.

[c]Although we illustrated the approach conceptually, this is in essence what is known as the *universal approximation theorem* [93,94].

DOI: 10.1201/9781003350101-8

Due to its structural similarity (see Fig. 8.2), this building block is known as an *artificial neuron* [95] and their arrangements are known as *neural networks*.

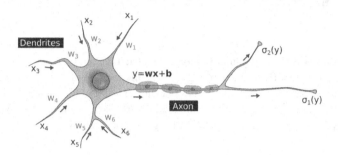

Figure 8.2 A biological neuron with similarities with an artificial neuron indicated. Inputs x_n are affinely transformed through the dendrites and activation is produced through the axon terminals

8.1 MCCULLOCH-PITTS ARTIFICIAL NEURON

When the activation function is a step function[d] and has a single output, our building block is referred to as McCulloch[e]-Pitts[f] [95] (MP) artificial neuron:

$$g(\mathbf{x}) = \Theta(\mathbf{w} \cdot \mathbf{x} + b). \tag{8.2}$$

Let's see how this neuron works by studying a few logic functions. In this case, our input is a vector with two elements, and we can write Eq. 8.2 as:

$$g(x_1, x_2) = \Theta(W_1 x_1 + W_2 x_2 + b). \tag{8.3}$$

For the OR function, we have that:

$$W_1 x_1 + W_2 x_2 + b \begin{cases} \geq 0 \text{ if } x_1 = 1 \lor x_2 = 1 \\ < 0 \text{ if } x_1 = 0 \land x_2 = 0, \end{cases} \tag{8.4}$$

which gives $W_n \geq -b$ and $b < 0$. For instance, we could choose $b = -1$ and set $W_1 = 1$, $W_2 = 1$, producing:

$$OR(x_1, x_2) = \Theta(x_1 + x_2 - 1). \tag{8.5}$$

From this equation, we can set a boundary that separates states 0 and 1, given by: $x_2 = 1 - x_1$. This is illustrated in Fig. 8.3.

---

[d]$\Theta(x) = 1 \iff x \geq 0$, 0 otherwise.

[e]Warren Sturgis McCulloch (1898–1969) American neurophysiologist, winner of the Wiener Gold Medal in 1968.

[f]Walter Harry Pitts, Jr. (1923–1969) American logician.

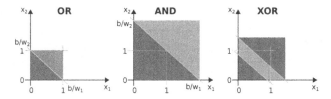

Figure 8.3 Implementations of AND and OR logic functions using an MP neuron. The XOR function is not linearly separable

For the AND function, we have that:

$$W_1 x_1 + W_2 x_2 + b \begin{cases} \geq 0 \text{ if } x_1 = 1 \wedge x_2 = 1 \\ < 0 \text{ if } x_1 = 0 \vee x_2 = 0, \end{cases} \quad (8.6)$$

which gives $W_n < -b$, $b < 0$ and $W_1 + W_2 \geq -b$. In this case, we could choose $b = -2$ and set $W_1 = W_2 = 1$, which produces:

$$AND(x_1, x_2) = \Theta(x_1 + x_2 - 2). \quad (8.7)$$

In this case, the boundary is given by $x_2 = 2 - x_1$ and is also illustrated in Fig. 8.3. Note, though, that this artificial neuron works as a linear classifier and an XOR function can not be obtained.

8.1.1 PERCEPTRON

Although we could successfully find the coefficients $\mathbf{w}$ analytically for some logical functions using the MP neuron, we need an automated procedure for doing this calculation. The *perceptron* [96] is one of the first supervised learning algorithms proposed to *train* an MP neuron.

Here, the idea is that, given a *training dataset*, a 0 misclassification implies that $\mathbf{w} \cdot \mathbf{x} + b \geq 0$ whereas a 1 misclassification implies that $\mathbf{w} \cdot \mathbf{x} + b < 0$. A compact way to write this error is $\varepsilon = \left(\hat{f}(\mathbf{x}) - f(\mathbf{x}) \right) (\mathbf{w} \cdot \mathbf{x} + b)$.

To apply the gradient descent method (see Sec. 6.5.) to the parameters $\mathbf{w}$ and b, we need the derivatives:

$$\frac{\partial \varepsilon}{\partial \mathbf{w}} = \left(\hat{f}(\mathbf{x}) - f(\mathbf{x}) \right) \mathbf{x}^T,$$
$$\frac{\partial \varepsilon}{\partial b} = \left(\hat{f}(\mathbf{x}) - f(\mathbf{x}) \right). \quad (8.8)$$

Using these results in the gradient method we get:

$$\mathbf{w}_{n+1} = \mathbf{w}_n - \eta \left(\hat{y}(\mathbf{x}) - f \right) \mathbf{x},$$
$$b_{n+1} = b_n - \eta \left(\hat{f}(\mathbf{x}) - f \right). \quad (8.9)$$

Thus, we arrive at the perceptron algorithm:

Algorithm 15: Perceptron

---

input: A training set of $\mathbf{x}, f(\mathbf{x})$ samples
▷ Initialize $\mathbf{w}$ and b with random values;
repeat
   ▷ Sample an $\mathbf{x}, f(\mathbf{x})$ pair from the training set;
   ▷ Calculate the output $\hat{f}(\mathbf{x}) = \Theta(\mathbf{w} \cdot \mathbf{x} + b)$;
   ▷ Update the parameters:
$$\mathbf{w}_{n+1} = \mathbf{w}_n - \eta(\hat{y}(\mathbf{x}) - f)\mathbf{x}$$
$$b_{n+1} = b_n - \eta(\hat{f}(\mathbf{x}) - f)$$
until *acceptable error*;

This algorithm is very simple to implement in Julia. First, we define a 4×3 matrix *example* whose rows have the two inputs and the desired output. We also define a learning rate $\eta = 0.001$:

```
while(ε > 0)
    ε = 0
    for epoch in 1:N
        r=rand(1:4)
        x=example[r,1:2]+rand(2)/100
        f=example[r,3]
        h_hat=Θ(w·x+b)

        Δf=h_hat-f
        w = w - η*Δf*x
        b = b - η*Δf
        ε += abs(Δf)
    end
    ε /= N
end
```

Note that i) we run the algorithm with N different examples to compute an average error, and ii) we add a small noise to the input to have variability of the training set.

8.2   MULTILAYER PERCEPTRON

Both the MP neuron and the perceptron algorithm can only work with linearly separable datasets. We can improve the performance of an artificial neuron with a few updates. Let's consider the XOR logical function. It can be decomposed into a few more fundamental operations for example $A \oplus B = A\bar{B} + \bar{A}B$. This suggests that using a few neurons connected in some specific topology may render the results we are after. Let's try a feedforward configuration with a hidden layer composed of two neurons as shown in Fig. 8.4 and different activation functions.

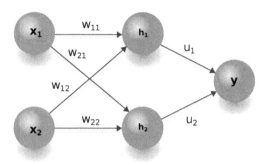

Figure 8.4   Implementation of an XOR function using artificial neurons

For this network[g], we have the following set of equations:

$$
\begin{aligned}
\mathbf{h} &= \sigma_h(\mathbf{Wx} + \mathbf{b}_h) \\
\hat{y} &= \sigma_y(\mathbf{u} \cdot \mathbf{h} + \mathbf{b}_y)
\end{aligned}
\tag{8.10}
$$

To find its parameters, we first need to define an error related to the output. Maybe the simplest errors one could think of would be L1 and L2 losses, given by $\|\hat{y} - y\|$ and $(\hat{y} - y)^2$ respectively. Since it is an analytical function, L2 loss is typically preferred. If the dataset contains outliers, though, errors can be amplified by the power function, and in these cases, the L1 error can be preferred. Let's use the L2 error for this example:

$$
L = (\hat{y} - y)^2.
\tag{8.11}
$$

It would be nice to apply the gradient descent algorithm, but for this, we need the derivative of the loss function with respect to each parameter and the derivative of the activation functions. Therefore, the step function used in the MP neuron cannot be used here because its derivative is not defined. Instead, we can use functions such as the sigmoid function if we want a unipolar output or the hyperbolic tangent if we want a bipolar output. Since we want to implement an XOR function, let's opt for the former for both layers.

---

[g]For references about the multilayer perceptron, see: [97, 98].

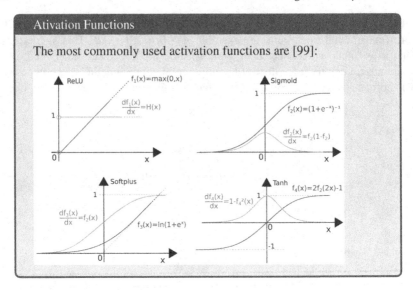

Ativation Functions

The most commonly used activation functions are [99]:

It is beneficial, though, to derive a more general approach for networks with any number of layers. In general, for an N-layered feedforward network, we have:

$$\mathbf{z}_1 = \sigma_1(\mathbf{x})$$

$$\vdots$$

$$\mathbf{z}_n = \sigma_n\left(\mathbf{W}_n \mathbf{z}_{n-1} + \mathbf{b}_n\right) \tag{8.12}$$

$$\vdots$$

$$\mathbf{z}_N = \sigma_N\left(\mathbf{z}_{N-1}\right)$$

and a loss that is a function of $\mathbf{z}_N$.

In Julia, we can create a mutable structure for a layer that contains all this information:

```julia
mutable struct DenseLayer
    x            :: Matrix
    W            :: Matrix
    b            :: Matrix
    z            :: Matrix
    activation   :: String
    ∂L∂W         :: Matrix
    ∂L∂b         :: Matrix
end
```

The neural network can be described by another mutable structure that contains its layers and loss function:

```
mutable struct NeuralNetwork
       Layers   :: Vector{DenseLayer}
       loss     :: String
end
```

To obtain an estimate from the neural network, we just propagate forward:

```
function FeedForward!(x, nn :: NeuralNetwork)
    z = x
    for L in nn.Layers
        z = forwardDense!(z,L)
    end

    return z
end
```

where the *forwardDense!* function just calculates the output of each layer and stores the result:

```
function forwardDense!(x,layer::DenseLayer)
    setfield!(layer,:x,x)
    y=layer.W*x+layer.b

    if (layer.activation == "tanh")
        z = tanh.(y)
    end

    setfield!(layer,:z,z)

    return z

end
```

To train the network, we can use the gradient descent technique, but for this, we need to compute some gradients [97, 98]. The gradient of the loss with respect to the bias $\mathbf{b}_n$ is given by:

$$\frac{\partial L}{\partial \mathbf{b}_n} = \left(\frac{\partial L}{\partial \mathbf{z}_n} \frac{\partial \mathbf{z}_n}{\partial \mathbf{y}_n} \right) \frac{\partial \mathbf{y}_n}{\partial \mathbf{b}_n}, \tag{8.13}$$

where

$$\mathbf{y}_n = \mathbf{W}_n \mathbf{z}_{n-1} + \mathbf{b}_n. \tag{8.14}$$

The term within parentheses is a row vector, whereas the last term is an identity matrix. Therefore, it can be written as:

$$\frac{\partial L}{\partial \mathbf{b}_n} = \frac{\partial L}{\partial \mathbf{z}_n} \frac{\partial \mathbf{z}_n}{\partial \mathbf{y}_n}. \tag{8.15}$$

For the weight matrix $\mathbf{W}_n$, we have:

$$\frac{\partial L}{\partial \mathbf{W}_n} = \frac{\partial L}{\partial \mathbf{z}_n} \frac{\partial \mathbf{z}_n}{\partial \mathbf{y}_n} \frac{\partial \mathbf{y}_n}{\partial \mathbf{W}_n} = \frac{\partial L}{\partial \mathbf{y}_n} \frac{\partial \mathbf{y}_n}{\partial \mathbf{W}_n}, \tag{8.16}$$

which is the contraction between a row vector and a rank-3 tensor. The latter is given by:

$$\frac{\partial (\mathbf{y}_n)_i}{\partial (\mathbf{W}_n)_{jk}} = \frac{\partial}{\partial (\mathbf{W}_n)_{jk}} \left[\sum_l (\mathbf{W}_n)_{il} (\mathbf{z}_{n-1})_l + (\mathbf{b}_n)_i \right]$$

$$= \sum_l \delta_{ij} \delta_{kl} (\mathbf{z}_{n-1})_l = \delta_{ij} (\mathbf{z}_{n-1})_k. \tag{8.17}$$

Plugging it back into the contraction:

$$\frac{\partial L}{(\partial \mathbf{W}_n)_{jk}} = \sum_i \left(\frac{\partial L}{\partial \mathbf{y}_n} \right)_i \frac{\partial (\mathbf{y}_n)_i}{\partial (\mathbf{W}_n)_{jk}}$$

$$= \sum_i \left(\frac{\partial L}{\partial \mathbf{y}_n} \right)_i \delta_{ij} (\mathbf{z}_{n-1})_k$$

$$\left(\frac{\partial L}{\partial \mathbf{W}_n} \right)_{kj} = \left(\frac{\partial L}{\partial \mathbf{y}_n} \right)_j (\mathbf{z}_{n-1})_k \tag{8.18}$$

$$\frac{\partial L}{\partial \mathbf{W}_n} = \mathbf{z}_{n-1} \frac{\partial L}{\partial \mathbf{y}_n}.$$

We must also calculate the term that backpropagates to the next layer:

$$\frac{\partial L}{\partial \mathbf{z}_{n-1}} = \frac{\partial L}{\partial \mathbf{z}_n} \frac{\partial \mathbf{z}_n}{\partial \mathbf{y}_n} \frac{\partial \mathbf{y}_n}{\partial \mathbf{z}_{n-1}} = \frac{\partial L}{\partial \mathbf{y}_n} \mathbf{W}_n. \tag{8.19}$$

Starting with the loss, we backpropagate through each layer using gradient descent to update their parameters until we reach the first layer where:

$$\frac{\partial L}{\partial \mathbf{W}_1} = \mathbf{x} \frac{\partial L}{\partial \mathbf{y}_1}. \tag{8.20}$$

In Julia, the backpropagation is very similar to the feedforward function, but backward over the layers:

```julia
function BackPropagate(∂L, nn :: NeuralNetwork)
    D = ∂L
    for L in nn.Layers[end:-1:1]
        D = backDense!(D,L)
    end

    return D
end
```

where the *backDense!* function computes the backpropagation through each layer:

```julia
function backDense!(∂L∂z,layer :: DenseLayer)
    if (layer.activation == "sig")
        D = (layer.z .* (1.0 .- layer.z))'
    end

    ∂L∂y = ∂L∂z.*D
    setfield!(layer,:∂L∂b,layer.∂L∂b + ∂L∂y)
    setfield!(layer,:∂L∂W,layer.∂L∂W + layer.x*∂L∂y)
    ∂L∂x = ∂L∂y*layer.W

    return ∂L∂x
end
```

Note that, instead of using numerical or symbolic differentiation, we can use explicit equations for the partial derivatives and backpropagate them. The term *differential programming* has been used to refer to such an approach (see Appendix D).

The training consists of feedforwarding, obtaining the loss and its gradient, and then backpropagating the gradient back through all layers. The final step is to just apply the gradients to the network:

```
function SetGradients!(η, nn :: NeuralNetwork)
    for L in nn.Layers
        Gradient!(L,η)
    end

    return nothing
end
```

where the function *Gradient!* applies the already computed gradients to each layer
and zeroes the cache:

```
function Gradient!(layer :: DenseLayer, η)
    setfield!(layer,:W, layer.W - η*layer.∂L∂W')
    setfield!(layer,:b, layer.b - η*layer.∂L∂b')

    i,j = size(layer.∂L∂W)
    z = zeros(i,j)
    setfield!(layer,:∂L∂W,z)

    i,j = size(layer.∂L∂b)
    z = zeros(i,j)
    setfield!(layer,:∂L∂b,z)

    return nothing
end
```

8.2.1 OPTIMIZATION STRATEGIES

Adjusting the learning rate can have a significant impact on training the network.
Consider Fig. 8.5, for instance. Small learning rates may render very slow trainings,
whereas high training rates may make the network diverge. Finding an adequate
learning rate is essential to perform a timely, yet reliable training of the network.

Also, we saw that many layers may be required to implement non-linear classifiers. A complication arises, though, if one uses backpropagation to train these layers.
The backpropagation requires products of gradients, and these, for many activation
functions, tend to be close to zero. Therefore, a nearly zero gradient may reach deep
layers and this may hinder proper learning. This is known as the *vanishing gradient
problem* [100]. One solution to this problem is to use different activation functions

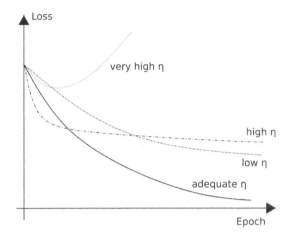

Figure 8.5   Effects on the learning rate η on the evolution of loss along the epoch

such as the rectified linear unit (ReLU), which has a constant derivative for positive values. Many other strategies have been proposed such as using non-gradient methods and skipping connections.

Conversely, using non-zero decaying gradients may produce the opposite effect, an *exploding gradient* [100]. This can be solved by applying an L2 regularization or *clipping* the gradient: $\nabla L \leftarrow G_{max} \nabla L / \|\nabla L\|$, where G_{max} is a threshold value.

Robbins-Monro Stochastic Approximation

Suppose you want to find the zeros of a function $g(\theta^*)$ but this function is not directly observable. Instead, you can observe a random variable y whose expected value is $g(\theta)$. Robbins and Monro[a] showed that it is still possible to find θ^* by using [101]:

$$\theta_{n+1} = \theta_n - \alpha_n y_n, \qquad (8.21)$$

as long as $\sum_n \alpha_n = \infty$ and $\sum_n \alpha_n^2 < \infty$.

This is known as the *Robbins-Monro stochastic approximation* method, which is the foundation of the *stochastic gradient descent* method that receives its name because of its similarity with the conventional gradient descent method (see [102] and Sec. 6.5.).

---

[a]Hebert Ellis Robbins (1915–2001), adviser of Sutton Monro (1919–1995), both American mathematicians.

When we backpropagate gradients for each random input sample, and update the network's parameters accordingly, this method is known as *stochastic gradient descent* (SGD) [101, 103]. However, samples may be very distinct from one another and this may not lead to very stable trainings. We can make the learning process more stable if we feed all samples to the network, take the average gradient and then backpropagate it. This is called *batch gradient descent* (BGD)[h]. Although it is more stable, it needs to process all samples before making an update and, consequently, has a longer convergence time. A third solution is to take a small set of input samples, feedforward them, calculate the average loss, and then backpropagate the average gradient. This last method is known as *mini-batch gradient descent* (MBGD). In this case, we have:

$$\langle f(\mathbf{x}) \rangle = \frac{1}{N} \sum_{i=1}^{N} f^i(\mathbf{x}), \tag{8.22}$$

where the superscript i indicates the position within the mini-batch. Thus, we have:

$$\begin{aligned} \mathbf{x}_{n+1} &= \mathbf{x}_n - \alpha [\nabla_x \langle f(\mathbf{x}) \rangle]^T \\ &= \mathbf{x}_n - \alpha \frac{1}{N_{MB}} \sum_{i=1}^{N_{MB}} \left[\nabla_x f^i(\mathbf{x}) \right]^T. \end{aligned} \tag{8.23}$$

Figure 8.6 compares the trajectories of the batch gradient descent, the mini-batch gradient descent, and the stochastic gradient descent toward a minimum. Although it is faster to compute a stochastic gradient descent, its trajectory toward a minimum is slow because of its intrinsic random motion. Batch gradient descent, on the other hand, takes longer to compute as all samples have to be processed before updating the gradients. However, it moves more systematically toward the minimum. Mini-batch gradient descent is a compromise between both strategies.

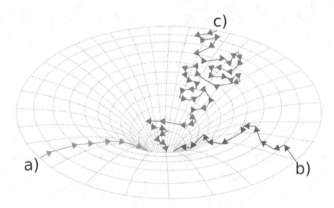

Figure 8.6 a) Batch gradient descent, b) Mini-batch gradient descent, and c) Stochastic gradient descent trajectories toward a minimum

The number of times it takes to show all the samples at least once to the network is called an *epoch*, and it takes several epochs to train the network. The difference between mini-batch and epoch is illustrated in Fig. 8.7.

---

[h]Not to be confused with the batch normalizing transform that is discussed in Appendix E.

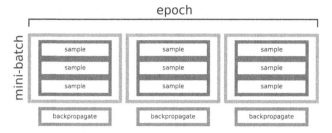

Figure 8.7   A total of nine samples organized in three mini-batches of three samples each

To verify how well the network can make generalizations, the data set is typically divided into a training set and a validation set. The former is used to train the network, whereas the latter is fed to the network once every epoch with the only purpose of checking its capacity to generalize. This data set is not used for backpropagation. If the validation loss is much higher than the training loss, then the network is not capable of making generalizations, which corresponds to a problem of overfitting. Overfitting can be originated, for example, if the training set has many correlations. Mini-batches are expected to contain uncorrelated samples, which helps train the network and minimize this problem.

Conversely, if the network produces lower validation loss than training loss, then the network is unable to properly train a model with this specific dataset, which corresponds to a problem of underfitting.

It can be difficult to train a neural network. We may want to have high learning rates to train the network quickly, but high learning rates may make cause divergence. On the other hand, if we choose a very low learning rate, it may take a long time to train the network as illustrated in Fig. 8.5. Next we discuss some strategies to circumvent these problems.

8.2.1.1   Scheduling

One strategy that can be used to circumvent problems with training is to use a learning rate schedule [102, 104]. For instance, it would be a good option, to begin with a high learning rate and then reduce it over the epochs. A simple scheduler can be computed using:

$$\eta_{n+1} = \max(\alpha\eta_n, \eta_{\min}),\qquad(8.24)$$

where n is the training epoch, $\alpha \in (0, 1)$ is a multiplicative factor, and $\eta_{\min}$ is the lower bound for the learning rate.

Alternatively, one can also use:

$$\eta = \frac{\eta_0}{\sqrt{t+1}},\qquad(8.25)$$

where η_0 is the learning rate at the beginning of training. Whereas the first alternative is linear, this later strategy decreases the learning rate smoothly.

One schedule policy that has proven empirically to give good results is [105]:

$$\eta_n = \begin{cases} \eta_T + \frac{1}{2}(\eta_0 - \eta_T)\left(1 + \cos\left(\frac{n\pi}{T}\right)\right), & \text{for } n < T \\ \eta_T, & \text{otherwise.} \end{cases} \quad (8.26)$$

where η_T is the target learning rate after a period T.

Using a high learning rate at the beginning of training may lead to instabilities in the network. It is possible, however, to *warm up* the network by starting with a low learning rate, gradually increasing it linearly up to a maximum value, and then applying a scheduling strategy to decrease it over subsequent epochs.

8.2.1.2 Momentum

There is no guarantee that the gradient methods will reach a global minimum. The network can actually get stuck in a local minimum. One idea to circumvent this problem is to borrow the concept of momentum [106] from physics. For a parameter $\mathbf{x}$:

$$\mathbf{x}_{n+1} = \mathbf{x}_n - \alpha\left[\nabla_x f(\mathbf{x}_n)\right]^T + \gamma\mathbf{v}_n, \quad (8.27)$$

where $\mathbf{v}$ contains the information of the previous update:

$$\mathbf{v}_n = \alpha\nabla_x f(\mathbf{x}_{n-1}) + \mathbf{v}_{n-1}. \quad (8.28)$$

A more convenient way to express the momentum optimizer is:

$$\begin{aligned} \mathbf{x}_{n+1} &= \mathbf{x}_n - \Delta_n \\ \Delta_n &= \alpha\left[\nabla_x f(\mathbf{x}_n)\right]^T + \gamma\Delta_{n-1}. \end{aligned} \quad (8.29)$$

8.2.1.3 AdaGrad

Let's write the gradient of a set of parameters $\theta = \begin{bmatrix} \theta_1 & \theta_2 & \dots & \theta_N \end{bmatrix}^T$ as:

$$\mathbf{g}^{(n)} = \left[\nabla_\theta f(\theta^{(n)})\right]^T. \quad (8.30)$$

This way, we can write the gradient descent method for this parameter as:

$$\theta^{(n+1)} = \theta^{(n)} - \eta\, \mathbf{g}^{(n)}. \quad (8.31)$$

Let's create a matrix:

$$\mathbf{G}^{(n)} = \begin{bmatrix} \sum_{m=1}^n g_1^{(n)^T} g_1^{(n)} & & \\ & \ddots & \\ & & \sum_{m=1}^n g_2^{(n)^T} g_2^{(n)} \end{bmatrix}. \quad (8.32)$$

The *Adaptive Gradient Descent* (AdaGrad) [107] method consists in updating the *learning rate* according to:

$$\theta^{(n+1)} = \theta^{(n)} - \frac{\eta}{\sqrt{\varepsilon I + \mathbf{G}^{(n)}}}\, \mathbf{g}^{(n)}. \quad (8.33)$$

8.2.1.4   AdaDelta

The problem with the AdaGrad algorithm is that it keeps summing square terms. Since the gradients being added are small, eventually, the sum becomes too small and the algorithm loses its power. To circumvent this problem, one can define:

$$\langle g^{(n)^2} \rangle = \gamma \langle g^{(n-1)^2} \rangle + (1-\gamma) g^{(n)^2}, \tag{8.34}$$

where $\gamma \sim 0.9$ typically. The *AdaDelta* algorithm, also known as *RMSprop*[i], is defined as:

$$\theta^{(n+1)} = \theta^{(n)} - \frac{\eta}{\sqrt{\varepsilon + \langle g^{(n)^2} \rangle}} g^{(n)} = \theta^{(n)} - \frac{\eta}{RMS[g^{(n)}]} g^{(n)}. \tag{8.35}$$

In all algorithms so far, including the current, the units do not match. The parameter is updated according to its gradient which has a different unit. Therefore, the learning rate has to incorporate the correct units. This can be circumvented by adopting Newton's method:

$$\theta^{(n+1)} = \theta^{(n)} - \frac{f'\left(\theta^{(n)}\right)}{f''\left(\theta^{(n)}\right)} \rightarrow \theta^{(n)} - \frac{RMS[\Delta\theta^{(n-1)}]}{RMS[g^{(n)}]} g^{(n)}, \tag{8.36}$$

where

$$RMS[\Delta\theta^{(n-1)}] = \sqrt{\varepsilon + \langle \Delta\theta^{(n-1)^2} \rangle}, \tag{8.37}$$

and

$$\langle \Delta\theta^{(n-1)^2} \rangle = \gamma \langle \Delta\theta^{(n-2)^2} \rangle + (1-\gamma)\Delta\theta^{(n-1)^2}. \tag{8.38}$$

8.2.1.5   Adam

In the *Adaptive Moment Estimation* (Adam) [108], we calculate averages of the past first two moments (mean and variance):

$$\begin{aligned} \mathbf{m}^{(n)} &= \beta_1 \mathbf{m}^{(n-1)} + (1-\beta_1)\mathbf{g}^{(n)} \\ \mathbf{v}^{(n)} &= \beta_2 \mathbf{v}^{(n-1)} + (1-\beta_2)\mathbf{g}^{(n)} \odot \mathbf{g}^{(n)}, \end{aligned} \tag{8.39}$$

where the β coefficients are close to 1.

Both $\mathbf{m}$ and $\mathbf{v}$ are biased toward 0. To correct this bias, we write:

$$\begin{aligned} \widehat{\mathbf{m}}^{(n)} &= \frac{\mathbf{m}^{(n)}}{1-\beta_1^n}, \\ \widehat{\mathbf{v}}^{(n)} &= \frac{\mathbf{v}^{(n)}}{1-\beta_2^n}. \end{aligned} \tag{8.40}$$

---

[i]Proposed by Geoffrey Hinton (1947–present) British psychologist, but never published as a paper.

The update rule is then written as:

$$\theta^{(n+1)} = \theta^{(n)} - \eta \widehat{\mathbf{m}}^{(n)} \oslash \left(\left(\widehat{\mathbf{v}}^{(n)} \right)^{\circ 1/2} + \varepsilon \mathbf{E} \right). \tag{8.41}$$

It is suggested that initial values for the parameters are $\beta_1 \sim 0.9$, $\beta_2 \sim 0.999$, and $\varepsilon \sim 10^{-8}$.

8.2.1.6 Implementation Overview

Fortunately, Julia [5] has the Flux library [109, 110], which simplifies the integration of all these solutions. For example, let's define a neural network architecture composed of two layers. The first layer has *inputChannels* inputs, 32 neurons, a tanh activation, and the parameters are initiated randomly according to the fan-in and -out of the layer[j]. The second layer has *outputChannels* and a sigmoid activation. For this, we write:

```
function SetNet(inputChannels,outputChannels)
    m = Chain(
        Dense(inputChannels,32,tanh; init = Flux.glorot_normal()),
        Dense(32,outputChannels,sigmoid; init = Flux.glorot_normal())
    )

    return m
end
```

We can attach an optimizer such as Adam to our model. For this, we can create a struct that encapsulates both the architecture and the optimizer. Let's do this and create a function that returns an approximator:

```
mutable struct NNApprox
    model :: Chain
    optimiser :: Flux.Optimiser
end

function SetAppr(η)
    model = SetNet(2,1)
    op = Flux.Optimiser(ClipValue(1E-1),Flux.Optimise.ADAM(η))

    return NNApprox(model,op)
end
```

---

[j]This is also known as the "Xavier" initialization [111]. "He" initialization [112] is referred to an initialization where only the fan-in is used.

Observe that we also included gradient clipping to prevent exploding gradients.

Next, we need to train our model. To do this, first, we create a data set of input values x and expected outputs y. Function *DataLoader* creates an iterable object over mini-batches, whereas function *ncycle* cycles an iterator over many epochs. Once we have this iterator over mini-batches and epochs, we need to compute gradients using *Flux.gradient* over the parameters of the model according to some loss. Finally, we apply this gradient using *Flux.Optimise.update!*:

```
train_data = Flux.Data.DataLoader((x,y),batchsize=batch_size,shuffle=true)
train_data = ncycle(train_data,epochs)

for (a,b) in train_data
    ∇ = Flux.gradient(params(ap.model)) do
        loss = Flux.Losses.mse(ap.model(a),b)
    end
    Flux.Optimise.update!(ap.optimiser,Flux.params(ap.model),∇)
end
```

8.3   AUTOENCODERS AND GENERATIVE MODELS

Autoencoders [113, 114] consist of two networks, one encoder E equipped with a mapping function: $f_\phi : \mathbb{R}^n \to \mathbb{R}^p$ with parameters ϕ, and a decoder D equipped with a mapping function $g_\theta : \mathbb{R}^p \to \mathbb{R}^n$ with parameters θ. Autoencoders also include a bottleneck layer between the encoder and decoder networks. This is given by $h : \mathbb{R}^p \to \mathbb{R}^p$ which is known as *latent space, feature space*, or *embedding*.

If the input dimension is bigger than the hidden dimension $n > p$, we say that the autoencoder is *undercomplete*. When the hidden layer has a size p bigger than n, we say that it is a *sparse autoencoder* and the autoencoder is said to be *overcomplete*.

Autoencoders can be used for performing a *data-specific learned compression* where the most relevant features are extracted in the latent space. Autoencoders can also be used to reduce the dimensionality of some information that has to go through a limited bandwidth channel such as an image being transmitted from one cell phone to another. The sender reduces the representation of the image, it is transmitted, and the receiver cell phone reconstructs the image with a decoder network. For this purpose, the network is trained with the same patterns for the input and the output, producing a loss function:

$$L(\mathbf{x}, \mathbf{r}) = \|\mathbf{x} - \mathbf{r}\|^2 = \|\mathbf{x} - g(f(\mathbf{x}))\|^2. \tag{8.42}$$

This approach can also be slightly modified to produce a noise-reduction autoencoder [115]. For this, some noise is included in the input but the original signal is expected at the output during training.

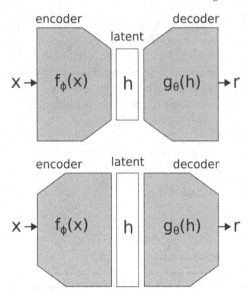

Figure 8.8   Autoencoder architecture with the encoder network, the hidden bottleneck, and the decoder network. Top: An undercomplete autoencoder architecture. Bottom: An overcomplete autoencoder architecture

To avoid overfitting in an autoencoder, it is possible to add a regularizer to the loss function:

$$L(\mathbf{x},\mathbf{r}) \mapsto L(\mathbf{x},\mathbf{r}) + \Omega(\mathbf{h}), \qquad (8.43)$$

where $\Omega(\mathbf{h})$ is a penalty function. This is common in *physics-informed machine learning* where physical conditions, such as energy conservation, are included in the penalty function. In other cases, it may be interesting, for example, to make the network less sensitive to small changes in the input. For this, we can use an L2 regularizer:

$$\Omega(\mathbf{h}) = \lambda \left\| \frac{\partial f_\phi(\mathbf{x})}{\partial \mathbf{x}} \right\|_F^2, \qquad (8.44)$$

where the F subscript implies a Frobenius[k] norm[1]. When using this type of regularizer, this network is usually known as a *contractive autoencoder*.

One use of autoencoders in physical science is to perform estimations of the orbits of objects. Let's consider the example below.

---

[k]Ferdinand Georg Frobenius (1849–1917) German mathematician, advised by Karl Weierstrass and adviser of Issai Schur, among others.

[1]$\|\mathbf{M}\|_F = \sqrt{\Sigma_{ij}|M_{ij}|^2}$.

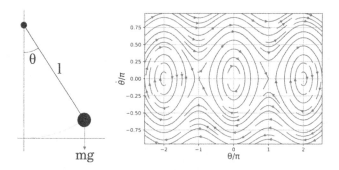

Figure 8.9 Left: Schematic of a simple pendulum, Right: Its phase space plotted using the function *streamplot* from PyPlot [2]

Example 8.3.1. Predicting the orbit of a simple pendulum.

The Hamiltonian* of a simple pendulum, shown in Fig. 8.9, is given by:

$$\mathcal{H} = \frac{p_\theta^2}{2ml} + mgl\,(1 - \cos(\theta)), \tag{8.45}$$

where p_θ and θ are the two canonical variables of the system.
Its equations of motion are given by:

$$
\begin{aligned}
\dot{\theta} &= \frac{\partial \mathcal{H}}{\partial p_\theta} = \frac{p_\theta}{ml^2} = \dot{\theta} \\
\dot{p}_\theta &= -\frac{\partial \mathcal{H}}{\partial \theta} \rightarrow \ddot{\theta} = -\omega_0^2 \sin(\theta),
\end{aligned} \tag{8.46}
$$

where $\omega_0^2 = g/l$ and we changed the canonical variable p_θ for $\dot{\theta}$.

With these equations, the phase space of the simple pendulum can be easily plotted as shown in Fig. 8.9. An overcomplete autoencoder with three hidden layers is enough to make good predictions of the trajectories of the pendulum in phase space. For instance, a network structure with 2 inputs ($\theta_i, \dot{\theta}_i$), hidden layers with 4, 8, and 4 neurons, and 2 outputs ($\theta_t, \dot{\theta}_t$) was trained to make the predictions shown in Fig. 8.10.

(*) William Rowan Hamilton (1805–1865) Irish mathematician and physicist winner of many prizes including the Royal Medal in 1835. ∎

8.3.1 VARIATIONAL AUTOENCODERS

When working with undercomplete autoencoders, the network may suffer from over-fitting. Consequently, the latent space may become irregular such that similar points may make the decoder produce very different results. Variational autoencoders [116] try to solve this problem by using a distribution of values as a latent space. In this technique, we use stochastic mappings for the encoder: $q_\phi(\mathbf{z}|\mathbf{x})$ with parameters ϕ,

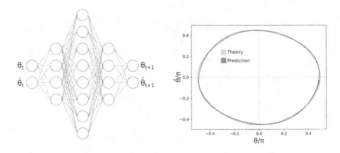

Figure 8.10 Left: Autoencoder model for predicting the trajectories of a simple pendulum in phase space, and Right: Prediction made by this autoencoder for an initial condition of $\theta_0 = \pi/2$ and $\dot{\theta}_0 = 0$

and the decoder: $p_\theta(\mathbf{x}|\mathbf{z})$ with parameters θ, as shown in Fig. 8.11. The network finds parameters of a distribution, which in statistics is known as *variational inference* [117].

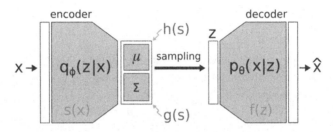

Figure 8.11 Stochastic autoencoder architecture

Let's assume that $q_\phi(\mathbf{z}|\mathbf{x})$ is Gaussian so that it can be described by a mean vector $\mu = g(\mathbf{x})$ and a covariance matrix $\Sigma = h(\mathbf{x})$. Therefore, for the distribution of $\mathbf{z}$, we can write a function $r_x(\mathbf{z}) \sim \mathcal{N}(g(\mathbf{x}), h(\mathbf{x}))$. The challenge is now to find functions g and h that produce a distribution $q_\phi(\mathbf{z}|\mathbf{x})$ close to our approximation $r_x(\mathbf{z})$. One way of doing this is minimizing the Kullback-Leibler divergence (see Sec. 2.2.1.) between them:

$$
\begin{aligned}
\mu, \Sigma &= \arg\min_{g,h} D_{KL}\left(r_x(\mathbf{z}) \| q_\phi(\mathbf{z}|\mathbf{x})\right) \\
&= \arg\min_{g,h} \left\langle \log r_x(\mathbf{z}) - \log q_\phi(\mathbf{z}|\mathbf{x}) \right\rangle.
\end{aligned}
\tag{8.47}
$$

Using Bayes theorem (see Sec. 2.1.2.1.):

$$
\begin{aligned}
\mu, \Sigma &= \arg\min_{g,h} \left\langle \log r_x(\mathbf{z}) - \log \frac{p_\theta(\mathbf{x}|\mathbf{z})q_\phi(\mathbf{z})}{p_\theta(\mathbf{x})} \right\rangle \\
&= \arg\max_{g,h} \left\langle \log p_\theta(\mathbf{x}|\mathbf{z}) - \log \frac{r_x(\mathbf{z})p_\theta(\mathbf{x})}{q_\phi(\mathbf{z})} \right\rangle.
\end{aligned}
\tag{8.48}
$$

We don't know the distribution of the latent space, so let's use $q_\phi(\mathbf{z})$ as a prior. We can use a simple prior such as the Gaussian distribution: $q_\phi(\mathbf{z}) \sim \mathcal{N}(\mathbf{0},\mathbf{I})$. Let's also use the conditional as a Gaussian distribution $p_\theta(\mathbf{x}|\mathbf{z}) = \mathcal{N}(f(\mathbf{z}),c\mathbf{I})$, where f is some function and c is a constant. Therefore, the posterior $q_\phi(\mathbf{z}|\mathbf{x})$ and the marginal $p_\theta(\mathbf{x})$ are also Gaussian, as assumed.

Thus, we arrive at:

$$\mu, \Sigma = \arg\max_{g,h} \left(\left\langle -\frac{|\mathbf{x} - f(\mathbf{z})|^2}{2c} \right\rangle - D_{KL}\left(r_x(\mathbf{z}) \| p_\theta(\mathbf{z})\right) \right). \tag{8.49}$$

Also, we seek a function $f(\mathbf{z})$ that maximizes the probability that $\hat{\mathbf{x}} = \mathbf{x}$ when $\mathbf{z}$ is sampled from an optimal $r_x(\mathbf{z})$. This will be the decoder part of the network. The encoder will be composed of a common core $s(\mathbf{x})$ and the two functions that approximate the mean ($\mu = g(s(\mathbf{x}))$) and the covariance ($\Sigma = h(s(\mathbf{x}))$). The encoder connects to the decoder through sampling, and this creates a problem for backpropagating the gradients. To circumvent this, it is possible to use a reparametrization trick where $\mathbf{z} = \mu + \Sigma^{1/2}\xi$, where $\xi \sim \mathcal{N}(\mathbf{0},\mathbf{I})$.

Connecting all components, we create four different networks (sx, h, g, and f) in Julia and propagate inputs forward with:

```
s = FeedForward!(x,sx)
μ = FeedForward!(s,h)
Σ = FeedForward!(s,g)
ξ = randn()
z = μ + sqrt(Σ)*ξ
xhat = FeedForward!(z,f)
```

The network loss is:

$$\begin{aligned}
\mathcal{L} &= \kappa \|\mathbf{x} - \hat{\mathbf{x}}\|^2 + D_{KL}\left(\mathcal{N}(\mu,\Sigma), \mathcal{N}(\mathbf{0},\mathbf{I})\right) \\
&= \mathcal{L}_1 + \mathcal{L}_2,
\end{aligned} \tag{8.50}$$

where $\kappa = 1/2c$, but it is more instructive to think of the loss as a regularized reconstruction error[m]. The first part is a typical L_2 loss for an autoencoder, whereas the second part is a regularization term for the latent space.

The backpropagation through the divergence is shown in Appendix F, and the backpropagation through the reparametrization gives:

$$\begin{aligned}
\frac{\partial \mathcal{L}}{\partial \mu} &= \frac{\partial \mathcal{L}_1}{\partial \mathbf{z}}\frac{\partial \mathbf{z}}{\partial \mu} + \frac{\partial \mathcal{L}_2}{\partial \mu} = \frac{\partial \mathcal{L}_1}{\partial \mathbf{z}} + \mu^\top, \\
\frac{\partial \mathcal{L}_1}{\partial \Sigma} &= \frac{\partial \mathcal{L}_1}{\partial \mathbf{z}}\frac{\partial \mathbf{z}}{\partial \Sigma} + \frac{\partial \mathcal{L}_2}{\partial \Sigma} = \frac{1}{2}\left(\xi\frac{\partial \mathcal{L}_1}{\partial \mathbf{z}}\right) \odot \Sigma^{-1/2} + 1/2\left(\mathbf{1} - \mathrm{inv}\left(\Sigma\right)^\top\right).
\end{aligned} \tag{8.51}$$

---

[m] This is often referred to as *evidence lower bound* (ELBO). See, for instance [117].

Attention must be taken to not have the regularization take over the training or the encoder may always produce values very close to $(\mathbf{0}, \mathbf{I})$. This is a phenomenon known as *posterior collapse*[n], where the latent variable is completely ignored, and the decoder is incapable of performing its job. The way to circumvent it is with a thermal annealing strategy (see Sec. 13.2) where the KL divergence is introduced slowly into the ELBO loss.

The backpropagation in Julia becomes:

```
∂L∂z = BackPropagate(∂L1,f)
∂L∂μ = ∂L∂z + ∂L2[1]*r
∂L∂Σ = 0.5*∂L∂z*ξ/sqrt(Σ) + ∂L2[2]*r
∂L∂s1 = BackPropagate(∂L∂μ,h)
∂L∂s2 = BackPropagate(∂L∂Σ,g)
∂L∂s = BackPropagate(∂L∂s1 + ∂L∂s2,sx)
```

8.3.1.1 Normalizing Flows

One restriction imposed in the construction of the variational autoencoder was the use of a Gaussian posterior. This is a simple distribution that might not capture the true distribution in many cases. Nonetheless, this problem can be alleviated using normalizing flows [119], a technique to produce more elaborated distributions through a series of invertible transformations.

The idea is to, given a random variable X, create a mapping to another random variable $Y = g(X)$. The probability of finding X within a differential element must be invariant under a change of variables because:

$$P[x < X < x + dx] = P[g(x) < g(X) < g(x + dx)].$$

Since we are dealing with a differential element, we can expand $g(x+dx)$ in a Taylor[o] series: $\approx g(x) + g'(x)dx$. Therefore:

$$P(x < X < x + dx) = P(y < Y < y + dy).$$

From these two last equations, we consequently have that $f_Y(y)|dy| = f_X(x)|dx|$. Now, given that $x = g^{-1}(y)$, we have:

$$dx = \frac{dg^{-1}(y)}{dy} dy$$

---

[n]See, for example [118].
[o]Brook Taylor (1685–1731) English mathematician.

and, consequently:

$$
\begin{aligned}
f_Y(y) &= \left| \frac{dx}{dy} \right| f_X(x) \\
&= \left| \frac{dg^{-1}(y)}{dy} \right| f_X\left(g^{-1}(y)\right).
\end{aligned}
\tag{8.52}
$$

If the distributions are multidimensional such that $\mathbf{y} = g(\mathbf{x})$, then:

$$
f_Y(\mathbf{y}) = f_X\left(g^{-1}(\mathbf{y})\right) |\det[\mathbf{J}]|,
\tag{8.53}
$$

where $\mathbf{J}$ is the Jacobian matrix of g^{-1} evaluated at $\mathbf{y}$.

Normalizing flows use this transformation to progressively change an original distribution f_0 to a more complex one through functions g_i applied to a sequence of random variables $\mathbf{z}_0, \mathbf{z}_1, \ldots, \mathbf{z}_k$. The transformation of probability densities is given by:

$$
f_i(\mathbf{z}_i) = f_{i-1}\left(g_i^{-1}(\mathbf{z}_i)\right) \left| \det\left[\frac{dg_i^{-1}}{d\mathbf{z}_i} \right] \right|.
\tag{8.54}
$$

Since $\det(\mathbf{M}^{-1}) = (\det(\mathbf{M}))^{-1}$ and for $y = f(x)$, $df^{-1}(y)/dy = (df(x)/dx)^{-1}$, we get:

$$
f_i(\mathbf{z}_i) = f_{i-1}(\mathbf{z}_{i-1}) \left| \det\left[\frac{dg_i}{d\mathbf{z}_{i-1}} \right] \right|^{-1}.
\tag{8.55}
$$

Taking the log, we get:

$$
\log f_k(\mathbf{z}_k) = \log f_0(\mathbf{z}_0) - \sum_{i=1}^{k} \log \left| \det\left[\frac{dg_i}{d\mathbf{z}_{i-1}} \right] \right|.
\tag{8.56}
$$

One example is given by the function $g(\mathbf{z}) = \mathbf{z} + \mathbf{u}h(\mathbf{w}^\top \mathbf{z} + b)$, where $\mathbf{u}$, $\mathbf{w}$, and b are hyperparameters. In this case, we have:

$$
\frac{dg}{d\mathbf{z}} = \mathbf{I} + \mathbf{u}^\top h'(\mathbf{w}^\top \mathbf{z} + b)\mathbf{w}
$$

and

$$
\det\left[\frac{dg}{d\mathbf{z}} \right] = 1 + \mathbf{u}^\top h'(\mathbf{w}^\top \mathbf{z} + b)\mathbf{w}.
$$

Thus, the flow for this case is given by:

$$
\log f_k(\mathbf{z}_k) = \log f_0(\mathbf{z}_0) - \sum_{i=1}^{k} \log \left| 1 + \mathbf{u}^\top h'(\mathbf{w}^\top \mathbf{z} + b)\mathbf{w} \right|.
\tag{8.57}
$$

8.3.2 DIFFUSION MODELS

A diffusion process [24, 120–122] is given by:

$$
dX_t = G[X_t, dt],
\tag{8.58}
$$

where X_t is a random variable, and $G[.]$ is a Markov propagator. This Itô[p] process can be described by the stochastic differential equation:

$$dX_t = \mu(X_t,t)dt + \sigma(X_t,t)dW_t, \tag{8.59}$$

where W_t is a Wiener[q] process such that $W_0 = 0$, $\langle W_t \rangle = 0$, and $W_t - W_s \sim \mathcal{N}(0,t-s)$ [24].

Let's take the drift term as $\mu(X_t,t) = -1/2\beta_t' X_t$ and the diffusion term as $\sigma(X_t,t) = \sqrt{\beta_t'}$, where β_t' is a variance schedule. In this situation, we have:

$$dX_t = -1/2\beta_t' X_t dt + \sqrt{\beta_t'}dW_t. \tag{8.60}$$

We are more interested in discrete processes, therefore we can apply the Euler[r]-Maruyama[s] method to this equation in an interval $[0,T]$ to obtain:

$$\begin{aligned} X_{n+1} &= X_n - 1/2\beta_{\tau_n}' X_n \Delta_t + \sqrt{\beta_{\tau_n}'}\Delta_W \\ &\approx \sqrt{1 - \beta_{\tau_n}'\Delta_t}X_n + \sqrt{\beta_{\tau_n}'\Delta_t}\eta_n, \end{aligned} \tag{8.61}$$

where $\eta \sim \mathcal{N}(0,1)$, $\Delta_t = T/N$ for N data points, and τ_n is the discretization of time within the interval. Function β', evaluated at point τ_n, is only scaled by the factor Δ_t. Therefore, we can create a discrete variance scale $\beta_n = \beta_{\tau_n}'\Delta_t$ and obtain:

$$X_n = \sqrt{1 - \beta_{n-1}}X_{n-1} + \sqrt{\beta_{n-1}}\eta_{n-1}. \tag{8.62}$$

This allows us to progressively apply noise to the original data. Given an initial condition, though, we can find X at any instant propagating it backward. For example, making $\alpha_n = 1 - \beta_{n-1}$ and propagating it backward once, we get:

$$X_n = \sqrt{\alpha_n}\left(\sqrt{\alpha_{n-1}}X_{n-2} + \sqrt{1-\alpha_{n-1}}\eta_{n-2}\right) + \sqrt{1-\alpha_n}\eta_{n-1}. \tag{8.63}$$

By recalling that the sum of variances characterizes the resulting distribution when adding Gaussian distributed variables, and extrapolating this process to its initial condition, we obtain:

$$X_n = \sqrt{\bar{\alpha}_n}X_0 + \sqrt{1-\bar{\alpha}_n}\eta_0, \tag{8.64}$$

where $\bar{\alpha}_n = \prod_{i=0}^{n}\alpha_i$. If $\alpha_n < 1$, then $\lim_{n\to\infty}\bar{\alpha}_n = 0$ and we end up only with noise.

Furthermore, the corresponding Fokker[t]-Planck[u] equation [24] is given by:

$$\frac{1}{2}\frac{\partial}{\partial X_t}\left(q(X_t)\beta_t' X_t\right) + \frac{1}{2}\frac{\partial^2}{\partial X_t^2}\left(q(X_t)\beta_t'\right) = \frac{\partial q(X_t)}{\partial t}$$
$$\frac{\partial^2}{\partial X_t^2}q(X_t) + X_t\frac{\partial}{\partial X_t}q(X_t) + q(X_t) = \frac{2}{\beta_t'}\frac{\partial q(X_t)}{\partial t}, \tag{8.65}$$

[p]Kiyosi Itô (1915–2008) Japanese mathematician, adviser of Shinzo Watanabe.
[q]Norbert Wiener (1894–1964) American mathematician.
[r]Leonhard Euler (1707–1783) Swiss polymath, advised by Johann Bernoulli.
[s]Gisiro Maruyama (1916–1986) Japanese mathematician.
[t]Adriaan Dániël Fokker (1887–1972) Dutch physicist, advised by Hendrik Lorentz.
[u]Max Karl Ernst Ludwig Planck (1858–1947) German physicist winner of many prizes, including the Nobel Prize in Physics in 1918. Planck was advised by Gustav Kirchhoff and Hermann von Helmholtz, among others. Planck advised many students, including Gustav Ludwig Hertz, Walter Schottky, and Max von Laue.

where $q(X_t)$ is the probability density of the random variable X_t. On the other hand, we know from Eq. 8.62 that the conditional probability is given by:

$$q(X_n|X_{n-1}) = \mathcal{N}(X_n; \sqrt{\alpha_n}X_{n-1}, \beta_n). \tag{8.66}$$

Also, from Eq. 8.64 we get:

$$q(X_n|X_0) = \mathcal{N}(X_n; \sqrt{\bar{\alpha}_n}X_0, 1 - \bar{\alpha}_n). \tag{8.67}$$

Now, the reverse diffusion process will have conditional probability:

$$p_\theta(X_{n-1}|X_n) = \mathcal{N}(X_{n-1}; \mu_\theta(X_n), \sigma_\theta(X_n)), \tag{8.68}$$

where μ_θ and σ_θ are learnable parameters. Instead of working with this conditional density, however, let's use Bayes' rule to obtain:

$$p_\theta(X_{n-1}|X_n, X_0) = q(X_n|X_{n-1}, X_0)\frac{q(X_{n-1}|X_0)}{q(X_n|X_0)}. \tag{8.69}$$

Using Eqs. 8.66, 8.67, and 8.64 we get:

$$p_\theta(X_{n-1}|X_n, X_0) \sim \mathcal{N}\left(X_{n-1}; \tilde{\mu}(X_n, X_0), \tilde{\beta}\right), \tag{8.70}$$

where

$$\begin{aligned}\tilde{\mu}(X_n, X_0) &= \frac{\sqrt{\bar{\alpha}_{n-1}}\beta_n}{1 - \bar{\alpha}_n}X_0 + \frac{\sqrt{\alpha_n}(1 - \bar{\alpha}_{n-1})}{1 - \bar{\alpha}_n}X_n \\ \tilde{\beta} &= \frac{(1 - \bar{\alpha}_{n-1})}{1 - \bar{\alpha}_n}\beta_n.\end{aligned} \tag{8.71}$$

According to Eq. 8.64, though, we can write $\tilde{\mu}$ as:

$$\tilde{\mu}(X_n, X_0) = \frac{1}{\sqrt{\alpha_n}}\left(X_n - \frac{1 - \alpha_n}{\sqrt{1 - \bar{\alpha}_n}}\eta_\theta(X_n)\right), \tag{8.72}$$

where η_θ is an estimator of the noise applied in the forward process. To find η_θ, we can variationally minimize the divergence between the forward and the backward distributions:

$$\begin{aligned}L &= D_{KL}\left(q(X_1, \ldots, X_T|X_0)\|p_\theta(X_0, \ldots, X_T)\right) \\ &= D_{KL}\left(\prod_{n=1}^{T}q(X_n|X_{n-1})\|p_\theta(X_T)\prod_{n=1}^{T}p_\theta(X_{n-1}|X_n)\right) \\ &= \left\langle \log\left(\frac{\prod_{n=1}^{T}q(X_n|X_{n-1})}{p_\theta(X_T)\prod_{n=1}^{T}p_\theta(X_{n-1}|X_n)}\right)\right\rangle_q \\ &= \left\langle -\log(p_\theta(X_T)) + \sum_{t=1}^{T}\log\left(\frac{q(X_n|X_{n-1})}{p_\theta(X_{n-1}|X_n)}\right)\right\rangle_q \\ &= \left\langle \log\left(\frac{q(X_T|X_0)}{p_\theta(X_T)}\right) + \sum_{n=2}^{T}\log\left(\frac{q(X_{n-1}|X_n, X_0)}{p_\theta(X_{n-1}|X_n)}\right) - \log p_\theta(X_0|X_1)\right\rangle_q \\ &= L_T + \sum_{n=2}^{T}L_{n-1} - L_0.\end{aligned} \tag{8.73}$$

The first term $L_T = D_{KL}(q(X_T|X_0)\|p_\theta(X_T))$ gives an error proportional to the deviance of X_T from noise but has no trainable parameters. L_0 is a reconstruction term that can be learned from a separate decoder. The remaining term $L_{n-1} = D_{KL}(q(X_{n-1}|X_n,X_0)\|p_\theta(X_{n-1}|X_n))$ is an error related to how the denoising backward step approximates the forward step. The divergence between two Gaussian distributions can be calculated analytically:

$$D_{KL}(q(X_{n-1}|X_n,X_0)\|p_\theta(X_{n-1}|X_n)) =$$
$$\frac{1}{2}\left[\log\left(\frac{|\Sigma_p|}{|\Sigma_q|}\right) + \mathrm{tr}\left(\Sigma_q^{-1}\Sigma_p\right) - d + (\mu_n - \tilde{\mu}(X_n))^\top \Sigma_q^{-1}(\mu_n - \tilde{\mu}(X_n))\right], \quad (8.74)$$

where d is the dimension of the distributions. Since we can make the variances of the distributions match exactly, the problem reduces to minimize the loss:

$$L_n = \frac{1}{2}\left[(\mu_n - \tilde{\mu}(X_n))^\top \Sigma_q^{-1}(\mu_n - \tilde{\mu}(X_n))\right]$$
$$= \frac{1}{2\|\Sigma(X_n)\|^2}\left\|\frac{1}{\sqrt{\alpha_n}}\left(X_n - \frac{1-\alpha_n}{\sqrt{1-\bar{\alpha}_n}}\eta_n\right)\right.$$
$$\left. - \frac{1}{\sqrt{\alpha_n}}\left(X_n - \frac{1-\alpha_n}{\sqrt{1-\bar{\alpha}_n}}\eta_\theta(X_n)\right)\right\|^2. \quad (8.75)$$

It has been found that training is better achieved when the weighting factor is neglected [?]. Thus, using Eq. 8.64 we get:

$$L_n = \left\|\eta - \eta_\theta\left(\sqrt{\bar{\alpha}_n}X_0 + \sqrt{1-\bar{\alpha}_n}\eta\right)\right\|^2. \quad (8.76)$$

Thus, we are looking for a model η_θ that learns how to infer the noise η applied to a signal at different diffusion stages. For training, we pick an X_0, $n \sim \mathrm{Uniform}(1,\ldots,N)$, add a noise signal η, and apply gradient descent on this loss function, as depicted in the algorithm below.

Algorithm 16: Training a diffusion model

---

input: X_0
repeat
  ▷ $n \sim \mathrm{Uniform}(\{1,\ldots,N\})$;
  ▷ $\eta \sim \mathcal{N}(\mathbf{0},\mathbf{I})$;
  ▷ Compute gradients for parameter θ on loss:
  $L_n = \left\|\eta - \eta_\theta\left(\sqrt{\bar{\alpha}_n}\mathbf{X}_0 + \sqrt{1-\bar{\alpha}_n}\eta\right)\right\|^2$, with $\bar{\alpha}_n = \prod_{i=1}^n(1-\beta_i)$;
until *convergence*;

For sampling, we begin with $X_T \sim \mathcal{N}(0,1)$ and, according to Eq. 8.72, consecutively apply, from $n = T,\ldots,1$:

$$X_{n-1} = \frac{1}{\sqrt{\alpha_n}}\left(X_n - \frac{1-\alpha_n}{\sqrt{1-\bar{\alpha}_n}}\eta_\theta(X_n)\right) + z, \quad (8.77)$$

where $z \sim \mathcal{N}(0,1)$ if $t > 1$ or 0 otherwise. This is summarized in the algorithm below:

Algorithm 17: Sampling from a diffusion model

---

input: $\mathbf{X}_N \sim \mathcal{N}(\mathbf{0}, \mathbf{I})$
for $n = N, \ldots, 1$ **do**
$\quad$ **if** $n > 1$ **then**
$\quad\quad \triangleright \mathbf{Z} \sim \mathcal{N}(\mathbf{0}, \mathbf{I})$;
$\quad$ **else**
$\quad\quad \triangleright \mathbf{Z} = \mathbf{0}$;
$\quad$ **end**
$\quad \triangleright \mathbf{X}_{n-1} = \frac{1}{\sqrt{\alpha_n}} \left(\mathbf{X}_n - \frac{1-\alpha_n}{\sqrt{1-\bar{\alpha}_n}} \eta_\theta(\mathbf{X}_n) \right) + \sigma_t \mathbf{Z}$;
end
return $\mathbf{X}_0$

Given that data is created through random inputs, this is known as a *generative model*.

8.3.2.1 Score-based Models

Another approach is to describe the reverse diffusion process by the stochastic differential equation [123]:

$$dX_t = \left[\mu(X_t,t) - \frac{1}{2}\sigma(X_t,t)^2 \nabla_X \log \, q_t(X_t) \right] dt + \sigma(X_t,t)d\tilde{W}_t, \qquad (8.78)$$

where q_t is the distribution of X_t in the forward process, and time flows backward. A network is trained by estimating a "score function" [124] (see Sec. 2.2.2.) $s_\theta(X_t)$ that minimizes $\|\nabla_X \log \, q_t(X_t) - s_\theta(X_t)\|^2$.

The Bayes' rule for a data set $X = x_1, \ldots, x_N$ gives:

$$\log \, p(\theta|X) = \log \, p(\theta) + \sum_{i=1}^{N} \log \, p(x_i|\theta) - \sum_{i=1}^{N} \log \, p(x_i). \qquad (8.79)$$

Applying gradient descent (see Sec. 6.5.), one gets:

$$\theta_{t+1} = \theta_t + \varepsilon \left[\nabla_\theta \log p(\theta) + \sum_{i=1}^{N} \nabla_\theta \log \, p(x_i|\theta) \right] + \eta_t, \qquad (8.80)$$

where ε is the training rate. Due to some resemblance with the Langevin[v] equation, wherein noise is injected, this processes is referred to as *Langevin dynamics*[w] [125]. This is useful because it only uses the score function, and, consequently, can be used to train score-based models.

---

[v]Paul Langevin (1872–1946) French physicist, advised by Pierre Currie and adviser of Louis de Broglie, among others. Winner of many prizes including the Hughes Medal in 1915.
[w]Not to be confused with the physical Langevin dynamics, which was derived by Paul Langevin.

8.3.3 GENERATIVE-ADVERSARIAL NETWORKS

The generative adversarial network [126] is another approach for generating synthetic data. It consists of two networks that compete against each other in a zero-sum game [24]. One network, the generator (G), generates data trying to fool the other network, the discriminator (D). The latter tries to estimate the probability that its input is real. As a result, the generator learns to produce realistic data with statistics close to that of the training set.

Since we have a classification problem, let's define a *value function* as the binary cross entropy:

$$V(G,D) = \langle \log(D(\mathbf{x})) \rangle + \langle \log(1 - D(G(\mathbf{z}))) \rangle, \qquad (8.81)$$

where $\mathbf{x}$ is real data, $\mathbf{z}$ is a random variable (latent[x] variable), $G(\mathbf{z})$ generates fake data, and $D(\mathbf{x})$ is the output of the discriminator as indicated in Fig. 8.12.

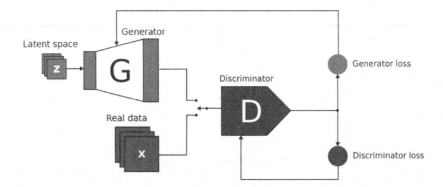

Figure 8.12   General structure of a GAN

When evaluating real data $\mathbf{x}$, we want $D(x)$ to produce the value 1, whereas when evaluating fake data, we want $D(G(\mathbf{z}))$ to produce the value 0. If the discriminator fails to identify real data, the value function becomes more negative. Therefore, the discriminator wants to maximize it. Conversely, the generator wants to produce $D(G(z)) = 0$, which corresponds to minimizing the value function. Hence, it is an adversarial network and the loss is often called *min-max* [127–129].

Let's define densities p_d and p_g for the discriminator and the generator so that we can rewrite the value function as:

$$V(G,D) = \int p_d(x) \log(D(x)) \, dx + \int p_g(x) \log(1 - D(G(z))) \, dx. \qquad (8.82)$$

The discriminator wants to maximize $V(G,D)$, and this happens when:

$$\frac{p_d(x)}{D(x)} - \frac{p_g(x)}{1 - D(x)} = 0 \rightarrow D^*(x) = \frac{p_d(x)}{p_d(x) + p_g(x)}. \qquad (8.83)$$

---

[x] A latent space here is defined as a space with lower dimensionality than that of the feature space.

For real data, we expect that $p_d(x) \to 1$, $p_g(x) \to 0$, which leads to $D^*(x) \to 1$. On the other hand, for fake data, we expect that $p_d(x) \to 0$, $p_g(x) \to 0$ leading to $D^*(x) \to 0$.

Using this result in the value function, we get:

$$
\begin{aligned}
V(G, D^*) &= \left\langle \log\left(\frac{1}{2}\frac{p_d(x)}{1/2\,[p_d(x)+p_g(x)])}\right)\right\rangle + \left\langle \log\left(\frac{1}{2}\frac{p_g(x)}{1/2\,[p_d(x)+p_g(x)]}\right)\right\rangle \\
&= -\log(4) + \left\langle \log(p_d(x)) - \log\left(\frac{p_d(x)+p_g(x)}{2}\right)\right\rangle \\
&\quad + \left\langle \log(p_g(x)) - \log\left(\frac{p_d(x)+p_g(x)}{2}\right)\right\rangle \\
&= -\log(4) + D_{KL}\left(p_d \,\|\, \frac{p_d+p_g}{2}\right) + D_{KL}\left(p_g \,\|\, \frac{p_d+p_g}{2}\right) \\
&= -\log(4) + D_{JS}(p_d \,\|\, p_g),
\end{aligned}
$$

(8.84)

where D_{KL} is the Kullback-Leibler divergence and D_{JS} is the Jensen-Shannon divergence (see Sec. 2.2.1.). This implies that the generator wants the JS divergence between the fake and real data to be the smallest possible ($p_d \to p_g$).

Training an adversarial network is usually not simple. Take, for instance, a value function $V(D, G) = xz$. Nash equilibrium [24] for this function, which corresponds to neither player being able to improve their objectives, happens when $x = z = 0$. Also, when computing gradients for optimization:

$$
\begin{aligned}
\Delta_x &= \frac{\partial xy}{\partial x} = y \\
\Delta_y &= \frac{\partial xy}{\partial y} = x,
\end{aligned}
$$

(8.85)

we see that they oscillate and increase in time. This implies that one network always countermeasures the other's strategy and some gradients will not produce an adequate training.

9 Advanced Network Architectures

In the previous chapter, we covered the basics of neural networks including the feedforward architecture and autoencoders. In this chapter, we will expand our discussion about neural networks by exploring some specialized architectures. We will begin with convolutional neural networks, which are tailored for image and computer vision applications. We will explore the mathematics behind this architecture and how they can be trained to perform advanced tasks such as object recognition. Following this discussion, we will investigate recurrent networks, which are typically used for processing sequential data, such as time series. Within the topic of recurrent networks, we will venture further into the attention mechanism, a tool that enables the algorithm to focus on specific elements of the data. This mechanism has been used in tasks such as machine translation. Moving on, we will delve into Hopfield and Boltzmann machines. Hopfield networks are a particular type of recurrent network that can store and retrieve patterns. The Boltzmann machine, on the other hand, is a powerful architecture that has been used in applications such as dimensionality reduction and generative modeling.

9.1 CONVOLUTIONAL NEURAL NETWORKS

Hubel[a] and Wiesel[b] have shown that the cerebral cortex of some animals contains neurons that respond to small regions of the visual field [130]. This discovery inspired the creation of convolutional models, such as the neocognitron [131,132], one of the first models for pattern recognition.

The convolution operation between two functions $x : \mathbb{R} \to \mathbb{R}$ and $w : \mathbb{R} \to \mathbb{R}$ is given by [133]:

$$(x \circledast w)(\tau) = \int_{-\infty}^{\infty} x(t)w(\tau - t)dt. \tag{9.1}$$

Its discrete version is described by:

$$
\begin{aligned}
(x \circledast w)_n &= \sum_{m=-M}^{M} x_m w_{n-m} \\
&= \sum_{m=-M}^{M} x_{n-m} w_m,
\end{aligned}
\tag{9.2}
$$

---

[a]David Hunter Hubel (1926–2013) Canadian neurophysiologist.
[b]Torsten Wiesel (1924–) Sweden neurophysiologist.

DOI: 10.1201/9781003350101-9

since the convolution is a commutative operation. In the equation, $\{-M,\ldots,M\}$ is a finite support[c].

The convolution can be written in a matrix format. Consider, for instance, a filter with three elements $\mathbf{w} = \begin{bmatrix} w_{-1} & w_0 & w_1 \end{bmatrix}^{\top}$ convolving it with a vector of three elements $\mathbf{x} = \begin{bmatrix} x_{-1} & x_0 & x_1 \end{bmatrix}^{\top}$. For this, we would obtain:

$$\begin{bmatrix} y_{-1} \\ y_0 \\ y_1 \end{bmatrix} = \begin{bmatrix} x_0 & x_{-1} & \\ x_1 & x_0 & x_{-1} \\ & x_1 & x_0 \end{bmatrix} \begin{bmatrix} w_{-1} \\ w_0 \\ w_1 \end{bmatrix} \rightarrow \mathbf{y} = \mathbf{T_x w}, \tag{9.3}$$

where $\mathbf{T_x}$ is a Toeplitz[d] matrix corresponding to the input $\mathbf{x}$.

Note that we are assuming that $x_{-2} = x_2 = 0$, which corresponds to padding the original vector $\mathbf{x}$ with zeros. This is known as *zero padding* or *same padding* [134]. Conversely, it is possible to only raster the filter inside the input array. This is known as *valid padding* and the resulting convolution will be smaller than the original input. Also, following the standard definition, we moved the filter one step at a time over the input signal, but, in principle, we could have used a different step. This is known as a *stride*. A comparison between different types of padding and stride is shown in Fig. 9.1.

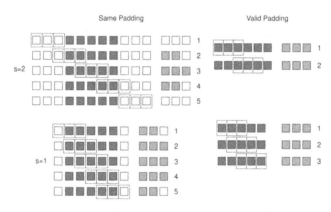

Figure 9.1   Same and valid paddings with strides 1 and 2

In general, the convolution operation produces an image with a new size:

$$N' = \left\lfloor \frac{N+p-k+s}{s} \right\rfloor, \tag{9.4}$$

where the symbol $\lfloor . \rfloor$ indicates the *floor*[e] rounding, p is the number of padding elements, k is the size of the kernel, and s is the stride. In our example, we have $N = 3$,

---

[c]The support of a measure μ is defined as the set Σ subtracted from the union of all open sets $E \in \Sigma$ whose measures are zero: $\sup(\mu) = \Sigma \backslash \{E \in \Sigma | \mu(E) = 0\}$, where $A \backslash B = \{x : x \in A \wedge x \notin B\}$ is the difference between sets.

[d]Otto Toeplitz (1881–1940) German mathematician.

[e]$\lfloor x \rfloor = \min\{a \in \mathbb{Z} | a \leq x\}$.

$p = 2$, $k = 3$, and $s = 1$. Therefore, the size of the output vector is 3. However, if we had used valid padding, then $p = 0$, and we would have obtained $N' = 1$.

It is common to convolve the signal with several filters and obtain multiple outputs corresponding to different properties of the input signal. For example, one filter might be an edge detector, while another can average the signal. This would result in outputs stacked as columns of $\mathbf{Y} = \mathbf{T_x W}$ with the columns of $\mathbf{W}$ being different filters if the input was still a vector.

In this case, we have for the nth output:

$$y_n = \sum_m \mathbf{x}_m \circledast \mathbf{w}_{m,n}. \tag{9.5}$$

where the inputs $\mathbf{x}_m$ are organized as columns of $\mathbf{X}$, and $\mathbf{W}$ is now a 3-dimensional array.

The convolution operation is applied along the rows, considering multiple inputs through the second dimension, and generating multiple outputs along the third dimension. Note that the second dimension of the kernel is exactly the number of input vectors and the third dimension of the kernel is exactly the number of desired outputs.

After convolution, a bias is added to the output, and the final result is passed through a non-linear activation function, typically a ReLU [135]:

$$\mathbf{y}_n \mapsto \sum_m \mathbf{x}_m \circledast \mathbf{w}_{m,n} + \mathbf{b}_n$$
$$\therefore \mathbf{z}_n = \sigma \left(\sum_m \mathbf{x}_m \circledast \mathbf{w}_{m,n} + \mathbf{b}_n \right). \tag{9.6}$$

After the convolution layer, the signal is shrunk through a process known as *pooling* [136]. The signal is divided into small segments and a new signal is created with the maximum or average of each segment. This is described for a 2D signal in Fig. 9.2.

When computing the gradient for *max pooling*, the maximum element of the segment receives the incoming gradient ∇L, while the remaining elements receive 0. For this purpose, we need to create a mask to use for backpropagation. For *average pooling*, all elements of the segment receive ∇L divided by the number of elements in the segment. Whereas maximum pooling extracts the most prominent features of a signal, average pooling smoothly extracts features.

Typically, in convolutional networks, the *softmax* activation is used in the last layer for identifying patterns. This activation is given by:

$$\hat{z}_i = \frac{e^{y_i - y_{\max}}}{\sum_j e^{y_j - y_{\max}}}, \tag{9.7}$$

where the term $y_{\max}$ is the maximum element of the vector $\mathbf{y}$ and assures that the exponentials are always bounded.

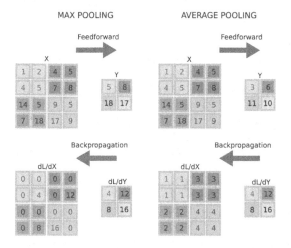

Figure 9.2 Max (left column) and average (right column) pooling

When performing pattern recognition, cross-entropy is typically used as a loss function. In this case, we have:

$$L = -\sum_i z_i \log(\hat{z}_i), \tag{9.8}$$

where $\mathbf{z}$ is a real distribution that can be used for training the neural network. If we combine this loss with the softmax activation, we get:

$$
\begin{aligned}
L &= -\sum_i z_i \left(y_i - \log \sum_j e^{y_j} \right) \\
&= -\sum_i z_i y_i + \log \sum_j e^{y_j}.
\end{aligned} \tag{9.9}
$$

and its gradient becomes:

$$\frac{\partial L}{\partial \mathbf{y}} = (\hat{\mathbf{z}} - \mathbf{z})^\top. \tag{9.10}$$

In Julia, we can use the library Flux [109, 110] to implement a convolutional neural network. The following model implements a convolutional neural network with three layers. The first does a $(1,1)$ padding of the input, performs a convolution with 16 $(3,3)$ filters, then applies a ReLU activation, and finally performs a max-pooling of the output. After the third layer, the output is flattened into an array and fed to a feedforward neural network with 10 outputs. Finally, a softmax output is computed to do pattern classification:

```
import Flux

model = Chain(
    Conv((3,3), 1=>16, pad=(1,1), relu)
    MaxPool((2,2)),

    Conv((3,3), 16=>32, pad=(1,1), relu)
    MaxPool((2,2)),

    Conv((3,3), 32=>32, pad=(1,1), relu)
    MaxPool((2,2)),

    flatten,
    Dense(288,10),
    softmax,
)
```

Training the network is performed as:

```
for (x,y) in training_set
    ∇ = Flux.gradient(Flux.params(model)) do
        ŷ = model(x)
        loss(ŷ,y)
    end

    Flux.Optimise.update!(ADAM(0.001),Flux.params(model),∇)
end
```

Flux.params returns the parameters of the model, loss is a function of the expected and estimated outputs y and $\hat{y}$, and *Flux.Optimise.update* updates the model with the Adam optimizer and the gradients computed with *Flux.gradient*. This network can be used to recognize digits, as depicted in Fig. 9.3.

9.2 RECURRENT NETWORKS

Recursive neural networks (RNNs) [106] can be understood as feedforward networks with internal states that behave as memory elements. Because of these internal states,

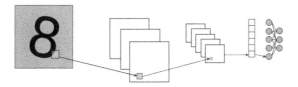

Figure 9.3 A pictorial example of a convolutional neural network that takes the image of a digit, performs convolutions, pooling, a flattening operation, and then feeds the signal to a dense layer

RNNs are Turing[f] complete and capable of simulating any Turing-computable function [137]. Given their general architecture, shown in Fig. 9.4, these networks are usually used for processing sequential data.

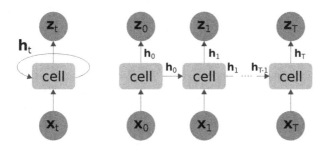

Figure 9.4 A recurrent neural network and its unfolding

The two simplest recurring configurations are the Elman[g] [138] and Jordan[h] [139] networks illustrated in Fig. 9.5 with a general behavior described by:

$$\begin{aligned} \mathbf{h}_t &= \sigma_h \left(\mathbf{W}_h \mathbf{x}_t + \mathbf{U}_h a(\mathbf{h}_{t-1}, \mathbf{z}_{t-1}) + \mathbf{b}_h \right) \\ \mathbf{z}_t &= \sigma_y \left(\mathbf{W}_y \mathbf{h}_t + \mathbf{b}_y \right). \end{aligned} \tag{9.11}$$

For Elman networks, $a(\mathbf{h}_{t-1}, \mathbf{z}_{t-1}) = \mathbf{h}_{t-1}$, whereas for Jordan networks, $a(\mathbf{h}_{t-1}, \mathbf{z}_{t-1}) = \mathbf{z}_{t-1}$.

Backpropagation in RNNs is known as *backpropagation through time* [140] and is a little trickier than that found in regular feedforward networks. Let's consider an Elman network:

$$\begin{aligned} \mathbf{h}_t &= \sigma_h(\mathbf{k}_t); \quad \mathbf{k}_t = \mathbf{W}_h \mathbf{x}_t + \mathbf{U}_h \mathbf{h}_{t-1} + \mathbf{b}_h \\ \hat{\mathbf{z}}_t &= \sigma_y(\mathbf{y}_t); \quad \mathbf{y}_t = \mathbf{W}_y \mathbf{h}_t + \mathbf{b}_y. \end{aligned} \tag{9.12}$$

---

[f] Alan Mathison Turing (1912–1954) English mathematician winner of the Smith's Prize in 1936, advised by Alonzo Church.

[g] Jeffrey Locke Elman (1948–2018) American cognitive scientist.

[h] Michael Irwin Jordan (1956–) American cognitive scientist, winner of many awards including the IEEE John von Neumann Medal in 2020.

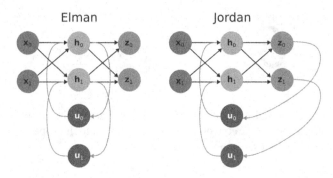

Figure 9.5 Elman (left) and Jordan (right) recurring neural network architectures

The total loss for this network is the average of all losses computed at each instant t:

$$L = \frac{1}{T} \sum_{t=1}^{T} l(\hat{\mathbf{z}}_t, \mathbf{z}_t). \tag{9.13}$$

The gradient of the loss with respect to $\theta \in \{\mathbf{W}_y, \mathbf{b}_y\}$ is then given by:

$$\nabla_\theta L = \frac{1}{T} \sum_{t=1}^{T} \frac{\partial l(\hat{\mathbf{z}}_t, \mathbf{z}_t)}{\partial \mathbf{y}_t} \frac{\partial \mathbf{y}_t}{\partial \theta}. \tag{9.14}$$

Therefore, we have:

$$\begin{aligned}
\frac{\partial L}{\partial \mathbf{b}_y} &= \frac{1}{T} \sum_{t=1}^{T} \frac{\partial l(\hat{\mathbf{z}}_t, \mathbf{z}_t)}{\partial \mathbf{y}_t} \\
\frac{\partial L}{\partial \mathbf{W}_y} &= \frac{1}{T} \sum_{t=1}^{T} \mathbf{h}_t \frac{\partial l(\hat{\mathbf{z}}_t, \mathbf{z}_t)}{\partial \mathbf{y}_t}.
\end{aligned} \tag{9.15}$$

These equations indicate that we can compute the loss for each time step individually, calculate their corresponding gradients, and then apply gradient descent using the sum of these gradients.

The gradient of the loss with respect to $\gamma \in \{\mathbf{W}_h, \mathbf{U}_h, \mathbf{b}_h\}$, on the other hand, is not so simple to compute since $\mathbf{h}_t$ depends on $\mathbf{h}_{t-1}$. To find it, we may rewrite Eq. 9.12 as:

$$\mathbf{h}_{t+1} = \sigma_h(\mathbf{W}_h \mathbf{x}_{t+1} + \mathbf{U}_h \mathbf{h}_t + \mathbf{b}_h) \tag{9.16}$$

and now take the gradient:

$$\begin{aligned}
\frac{d L}{d \mathbf{h}_t} &= \frac{d L}{d \mathbf{h}_{t+1}} \frac{\partial \mathbf{h}_{t+1}}{\partial \mathbf{h}_t} + \frac{\partial l(\hat{\mathbf{z}}_t, \mathbf{z}_t)}{\partial \mathbf{h}_t} \\
&= \frac{d L}{d \mathbf{h}_{t+1}} \sigma_h'(\mathbf{W}_h \mathbf{x}_{t+1} + \mathbf{U}_h \mathbf{h}_t + \mathbf{b}_h) \mathbf{U}_h + \frac{\partial l(\hat{\mathbf{z}}_t, \mathbf{z}_t)}{\partial \mathbf{y}_t} \frac{\partial \mathbf{y}_t}{\partial \mathbf{h}_t} \\
&= \frac{d L}{d \mathbf{h}_{t+1}} \sigma_h'(\mathbf{k}_{t+1}) \mathbf{U}_h + \frac{\partial l(\hat{\mathbf{z}}_t, \mathbf{z}_t)}{\partial \mathbf{y}_t} \mathbf{W}_y.
\end{aligned} \tag{9.17}$$

This is basically the gradient of the loss at the time instant t added of a term that comes from the next time step. This allows us to backpropagate through time and collect a set of gradients that can be added to update the parameters γ:

$$\begin{aligned}
\nabla_\gamma L &= \sum_{t=T}^{1} \frac{dL}{d\mathbf{h}_t} \frac{\partial \mathbf{h}_t}{\partial \mathbf{k}_t} \frac{\partial \mathbf{k}_t}{\gamma} \\
&= \sum_{t=T}^{1} \frac{dL}{d\mathbf{h}_t} \sigma'(\mathbf{k}_t) \frac{\partial \mathbf{k}_t}{\gamma} \\
&= \sum_{t=T}^{1} \frac{dL}{d\mathbf{k}_t} \frac{\partial \mathbf{k}_t}{\gamma},
\end{aligned} \tag{9.18}$$

where $dL/d\mathbf{k}_t = (dL/d\mathbf{h}_t)\sigma'(\mathbf{k}_t)$. Therefore:

$$\begin{aligned}
\frac{\partial L}{\partial \mathbf{b}_h} &= \sum_{t=T}^{1} \frac{dL}{d\mathbf{k}_t} \\
\frac{\partial L}{\partial \mathbf{W}_h} &= \sum_{t=T}^{1} \frac{dL}{d\mathbf{k}_t} \frac{\partial \mathbf{k}_t}{\partial \mathbf{W}_h} = \sum_{t=T}^{1} \mathbf{x}_t \frac{dL}{d\mathbf{k}_t} \\
\frac{\partial L}{\partial \mathbf{U}_h} &= \sum_{t=T}^{1} \frac{dL}{d\mathbf{k}_t} \frac{\partial \mathbf{k}_t}{\partial \mathbf{U}_h} = \sum_{t=T}^{1} \mathbf{h}_{t-1} \frac{dL}{d\mathbf{k}_y}.
\end{aligned} \tag{9.19}$$

Depending on how long back in time we need to backpropagate, this can be unfeasible. The product of gradients may blow up and even chaotic effects may appear with small changes in the initial conditions producing large outcomes. Therefore, this equation is often approximated by truncating it after a few steps, which is equivalent to directing the network's focus toward short-term effects.

The long short-term memory (LSTM) [141], shown in Fig. 9.6, is a network designed to avoid the vanishing gradient problem using "forget gates":

$$\begin{aligned}
\mathbf{f}_t &= \sigma\left(\mathbf{W}_{xf}\mathbf{x}_t + \mathbf{W}_{hf}\mathbf{h}_{t-1} + \mathbf{b}_f\right) \\
\mathbf{i}_t &= \sigma\left(\mathbf{W}_{xi}\mathbf{x}_t + \mathbf{W}_{hi}\mathbf{h}_{t-1} + \mathbf{b}_i\right) \\
\mathbf{g}_t &= \tanh\left(\mathbf{W}_{xc}\mathbf{x}_t + \mathbf{W}_{hc}\mathbf{h}_{t-1} + \mathbf{b}_c\right) \\
\mathbf{c}_t &= \mathbf{f}_t \odot \mathbf{c}_{t-1} + \mathbf{i}_t \odot \mathbf{g}_t \\
\mathbf{o}_t &= \sigma\left(\mathbf{W}_{xo}\mathbf{x}_t + \mathbf{W}_{ho}\mathbf{h}_{t-1} + \mathbf{b}_o\right) \\
\mathbf{h}_t &= \mathbf{o}_t \odot \tanh\left(\mathbf{c}_t\right) \\
\mathbf{z}_t &= \text{softmax}\left(\mathbf{W}_{hz}\mathbf{h}_t + \mathbf{b}_z\right).
\end{aligned}$$

Since the LSTM has a larger number of gates and, consequently, equations, its operation and training are more elaborated.

9.2.1 ATTENTION MECHANISM

One of the applications of the recurrent neural network is as a translator of sequences [142, 143]. This, however, can be easier implemented using attention

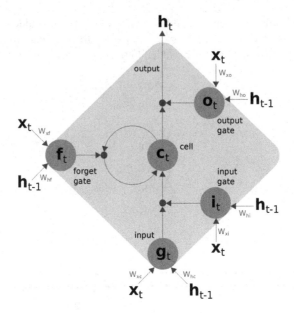

Figure 9.6   Architecture of an LSTM network

mechanisms [144, 145] that allow the network to focus on specific parts of a sequence. For example, in the sentence: "a bird flies high" it may be interesting to first
pay attention to the fact that we are talking about a bird. Let's embed these words in
vectors:

$$\mathbf{k}_1 = \begin{bmatrix} 0 & 0 & 0 \end{bmatrix}^\top$$
$$\mathbf{k}_2 = \begin{bmatrix} 1 & 1 & 2 \end{bmatrix}^\top$$
$$\mathbf{k}_3 = \begin{bmatrix} -1 & -1 & -2 \end{bmatrix}^\top$$
$$\mathbf{k}_4 = \begin{bmatrix} 0 & -1 & 3 \end{bmatrix}^\top$$

respectively. These vectors compose the matrix $\mathbf{K}$ of keys:

$$\mathbf{K} = \begin{bmatrix} \mathbf{k}_1 & \mathbf{k}_2 & \mathbf{k}_3 & \mathbf{k}_4 \end{bmatrix}$$
$$= \begin{bmatrix} 0 & 1 & -1 & 0 \\ 0 & 1 & -1 & -1 \\ 0 & 2 & -2 & 3 \end{bmatrix}.$$

We can query the word "bird" ($\mathbf{q} = \mathbf{k}_2$) by computing "energies"[i]:

$$\mathbf{e} = \mathbf{q}^\top \mathbf{K} = \begin{bmatrix} 0 & 6 & -6 & 5 \end{bmatrix}. \tag{9.20}$$

---

[i]It is common to normalize it by the square of the dimension of the keys.

The alignment of the query with the keys is then given by an *alignment score function* such as the cosine similarity of the softmax activation:

$$\mathbf{a} = \text{softmax}(\mathbf{e}) = \begin{bmatrix} 0.002 & 0.730 & 0.000 & 0.268 \end{bmatrix}. \tag{9.21}$$

When the elements of the softmax activation are multiplied by an inverse temperature such that the result is one at the position of the highest score and zero at all other positions, this is then called *hard* attention.

Now, suppose we want to give each of the keys an indication of how well they translate to the Portuguese word *penas* (feathers). We then create a vector of values:

$$\mathbf{v} = \begin{bmatrix} 0 & 0.7 & 0.4 & 0.1 \end{bmatrix}^{\top}.$$

With this, we can estimate how well our query (bird) translates to "penas" by weight-summing the energies:

$$s = \mathbf{a} \cdot \mathbf{v} = 0.54. \tag{9.22}$$

This is often called *context*.

Typically, this is used as a layer in neural network architectures, and for such, a batched input $\mathbf{X} = \begin{bmatrix} \mathbf{x}_1 & \dots & \mathbf{x}_N \end{bmatrix}$ is multiplied by trainable tensors $\mathbf{W}^Q$, $\mathbf{W}^K$, and $\mathbf{W}^V$ to produce multidimensional arrays:

$$\begin{aligned} \mathbf{Q} &= \mathbf{X}^{\top} \mathbf{W}^Q \\ \mathbf{K} &= \mathbf{X}^{\top} \mathbf{W}^K \\ \mathbf{V} &= \mathbf{X}^{\top} \mathbf{W}^V, \end{aligned} \tag{9.23}$$

respectively. This results in a context vector, instead of a scalar:

$$\mathbf{s} = \text{softmax}(\mathbf{Q}\mathbf{K}^{\top})\mathbf{V}. \tag{9.24}$$

When the query, the keys, and the values are equal, we call it *self-attention*.

In typical applications, many sets of multidimensional arrays $\mathbf{Q}_n$, $\mathbf{K}_n$, and $\mathbf{V}_n$ are used to produce different *heads* that are then averaged out by another learnable tensor. This is known as *multi-head attention*:

$$\mathbf{Z} = (\mathbf{s}_1 \| \mathbf{s}_2 \| \dots \| \mathbf{s}_N) \, \mathbf{T}. \tag{9.25}$$

Words, however, depend on their position in a sentence. To include this information in the model, a positional encoding vector is summed to the input embeddings. This can be a linear vector with values between 0 and 1, or more sophisticated functions including the sine and cosine [145]:

$$\begin{aligned} P(pos, 2i) &= \sin\left(\frac{pos}{(10^4)^{2i/d}}\right) \\ P(pos, 2i+1) &= \cos\left(\frac{pos}{(10^4)^{2i/d}}\right), \end{aligned} \tag{9.26}$$

where d is the dimension of the input, and i is the element number in the array.

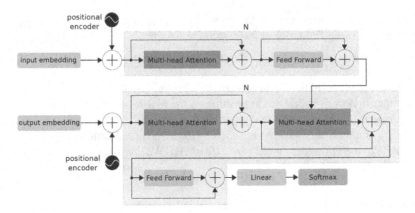

Figure 9.7 Transformer network using attention mechanisms

The attention mechanism is used in a transformer network, illustrated in Fig. 9.7 [145]. One of the applications of this network is in natural language processing where the inputs can be a text in one language (A) and the output can be a text in another language (B). The transformer network is composed of an encoder with N blocks consisting of a multi-head attention mechanism and a feed-forward layer. The latter is used to process the attention information produced by the previous layer. Also, note that this block uses *skip connections* that are used to amend the vanishing gradient over so many layers [146]. The decoding block consists of two multi-head attention layers and a feed-forward layer. The outputs, which are words from language B translated from language A, are sequentially fed into the decoder, shifting one word at a time to the right. The decoder then calculates the probability of each translation using the softmax activation.

9.3 HOPFIELD NETWORK

If the following sentence is given: "I think, therefore...," you probably guessed that the next two words would be "I am" because you have been exposed to it many times. The capability of recovering a pattern given a partial description is called *autoassociative memory* [147]. To understand how it works, let's briefly study how biological neurons can be mathematically modeled.

One of the earliest mathematical models for a biological neuron is the Hodgkin[j]-Huxley[k] model [148]. Consider the circuit shown in Fig. 9.8 with the switch connecting the current source. In a simplified Hodgkin-Huxley model, called *leaky integrate-and-fire* [149], the capacitor represents the capacitance of the cell membrane, the resistor describes ion channels, the battery depicts the electrochemical gradient that

---

[j] Alan Lloyd Hodgkin (1914–1998) English biophysicist, winner of many prizes including the Nobel Prize in Physiology in 1963.

[k] Andrew Huxley (1917–2012) English biophysicist winner of many prizes including the Nobel prize in Physiology share with A. Hodgkin in 1963.

drives the ion flow through the cell, and the current source models ion pumps. The voltage $V(t)$ across the membrane is described by the circuit law:

$$C\frac{dV(t)}{dt} = I(t) - \frac{V(t) - V_0}{R}. \tag{9.27}$$

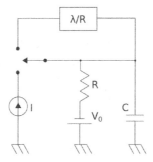

Figure 9.8 Circuital representation of a biological neuron

Consider now an autapse, where the neuron is self-excited. This is modeled by the switch now connecting the feedback loop in Fig. 9.8. In this case, the circuit equation becomes:

$$\tau\frac{dV(t)}{dt} = (\lambda - 1)V(t) - V_0, \tag{9.28}$$

where $\tau = RC$.

If the neuron is in equilibrium with the medium such that $V(0) = -V_0$, the evolution of the membrane voltage is given by:

$$V(t) = -\frac{V_0}{\lambda - 1}\left(\lambda \exp\left\{(\lambda - 1)\frac{t}{\tau}\right\} - 1\right). \tag{9.29}$$

It is simple to see that if $\lambda < 1$, the neuron relaxes to $V_0/(\lambda - 1)$, while if $\lambda > 1$, it grows exponentially toward $-\infty$. These are two states that correspond to attractors of the system. This suggests that, instead of using the adjustable parameter λ, we can use a function that saturates at two different values, and they must be determined by the current state of the system. Therefore, we can use an affine map as the input, the same way we did it in Chapter 8. Based on these arguments, for a vector state $\mathbf{u}(t)$, we can write:

$$\tau\frac{d\mathbf{u}(t)}{dt} = -\mathbf{u}(t) + \sigma\left(\mathbf{W}\mathbf{u}(t) + \mathbf{b}\right), \tag{9.30}$$

where $\mathbf{W}$ is a weight matrix, $\mathbf{b}$ is a bias vector, and σ is an activation function.

Let's discretize this equation, assume $\tau \approx \Delta_t$, and use a sign[l] activation function. In this case, we get:

$$\mathbf{u}_{n+1} = \text{sgn}\left(\mathbf{W}\mathbf{u}_n + \mathbf{b}\right). \tag{9.31}$$

This is the basic concept of a Hopfield[m] network [147].

---

[l]The sign function is given by: $\text{sgn}(x) = \begin{cases} +1 & \text{if } x > 0 \\ -1 & \text{otherwise.} \end{cases}$

[m]John Joseph Hopfield (1933–) American physicist.

Ideally, we want the network to remember certain patterns, that we shall call $\mathbf{x}|x_i \in [-1, +1]$, $i = 1, \ldots, N$:

$$\mathbf{u}_{n+1} = \text{sgn}(\mathbf{Wx} + \mathbf{b}) = \mathbf{x}. \tag{9.32}$$

For the weight matrix, one could try to use a simple Hebb[n]'s rule[o] [151] $\mathbf{W} = \eta \mathbf{xx}^T$, $\eta > 0$. However, we must avoid individual self-loops. Therefore, we can try $\mathbf{W} = \eta \left(\mathbf{xx}^T - \mathbf{I} \right)$:

$$
\begin{aligned}
\mathbf{u}_{n+1} &= \text{sgn}\left(\eta \left(\mathbf{xx}^T - \mathbf{I} \right) \mathbf{x} + \mathbf{b} \right) \\
&= \text{sgn}\left(\eta \left(\mathbf{xx}^T \mathbf{x} - \mathbf{x} \right) + \mathbf{b} \right) \\
&= \text{sgn}\left(\eta \left(\mathbf{x}N - \mathbf{x} \right) + \mathbf{b} \right) \\
&= \text{sgn}\left((N - 1)\eta \mathbf{x} + \mathbf{b} \right).
\end{aligned}
\tag{9.33}
$$

We could make $\eta = 1/(N - 1)$ to get $\mathbf{u}_{n+1} = \text{sgn}(\mathbf{x} + \mathbf{b})$, but anyway, the sign function is invariant under a positive multiplication of its argument. Hence, the weight matrix:

$$\mathbf{W} = \mathbf{xx}^T - \mathbf{I} \tag{9.34}$$

enables this dynamical system to "remember" the input pattern. This behavior can easily be visualized by considering $\mathbf{b} = \mathbf{0}$.

Thus, a Hopfield network can be understood as a fully connected bidirectional graph with symmetric weights and without self-loops where the edges are MP neurons with bipolar inputs and outputs as shown in Fig. 9.9. The network operates by setting an initial state on the nodes of the network and then allowing them to evolve synchronously or asynchronously over time. Asynchronous updates provide a closer representation of the biological process observed in the human brain.

9.3.1 TRAINING

For the Hopfield network shown in Fig. 9.9, the weight matrix is:

$$
\mathbf{W} = \begin{bmatrix} 0 & W_{12} & W_{13} \\ W_{12} & 0 & W_{23} \\ W_{13} & W_{23} & 0 \end{bmatrix}. \tag{9.35}
$$

Note that, since there are no self-loops, the diagonal elements are zero.

For a single training pattern $\mathbf{x}|x_i \in [-1, 1]$, $i = 1, \ldots, N$ for a Hopfield network consisting of N nodes, the weights are obtained via Hebb's rule, as previously stated. Since there are no self-loops on Hopfield networks, we must ensure that the diagonal elements of the weight matrix are zero by doing:

$$W_{ij} = x_i x_j - \delta_{ij} \rightarrow \mathbf{W} = \mathbf{xx}^T - \mathbf{I}. \tag{9.36}$$

---

[n]Donald Olding Hebb (1904–1985) Canadian psychologist.
[o]The Hebbian theory is often summarized and simplified as "neurons that fire together wire together" [150].

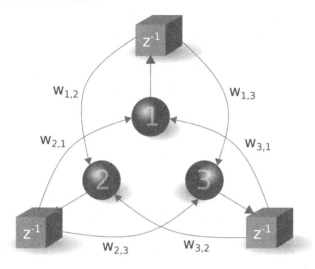

Figure 9.9 Example of a Hopfield network with three neurons and delay elements z^{-1}

If M training patterns $\{x_i\}_{i=1}^{M}$ are used, we get:

$$\mathbf{W} = \left\langle \mathbf{x}\mathbf{x}^T - \mathbf{I} \right\rangle = \frac{1}{M}\sum_{i=1}^{M}\mathbf{x}_i\mathbf{x}_i^T - \mathbf{I}. \tag{9.37}$$

For instance, let's consider the training patterns $x_1 = [-1, 1, -1]$, $x_2 = [-1, -1, 1]$, and $x_3 = [1, -1, -1]$. Applying Hebb's rule we get:

$$
\mathbf{W} = \frac{1}{3}\left(\begin{bmatrix} -1 \\ 1 \\ -1 \end{bmatrix}\begin{bmatrix} -1 \\ 1 \\ -1 \end{bmatrix}^T + \begin{bmatrix} -1 \\ -1 \\ 1 \end{bmatrix}\begin{bmatrix} -1 \\ -1 \\ 1 \end{bmatrix}^T + \begin{bmatrix} 1 \\ -1 \\ -1 \end{bmatrix}\begin{bmatrix} 1 \\ -1 \\ -1 \end{bmatrix}^T \right) - \begin{bmatrix} 1 & 0 & 0 \\ 0 & 1 & 0 \\ 0 & 0 & 1 \end{bmatrix}
$$

$$
= \begin{bmatrix} 0 & -1/3 & -1/3 \\ -1/3 & 0 & -1/3 \\ -1/3 & -1/3 & 0 \end{bmatrix}.
$$

$$\tag{9.38}$$

Hebb training of the Hopfield network in Julia is performed with the snippet below. Here, *patterns* is a matrix whose columns are the patterns that we want to store:

```
W = zeros(N,N)
for x in eachcol(patterns)
    W += x*transpose(x) - I
end
W /= size(patterns, 1)
```

The update rule for the state s_i of a Hopfield network is given by:

$$s_i^{t+1} = \begin{cases} +1 & \text{if } \sum_j W_{ij} s_j^t \ge b_i, \\ -1 & \text{otherwise.} \end{cases} \tag{9.39}$$

where b_i is a threshold value (bias) for the ith unit, which is typically chosen to be 0. In matrix form, the update rule can be written as:

$$\mathbf{s}^{t+1} = \operatorname{sgn}\left(\mathbf{W}\mathbf{s}^t + \mathbf{b}\right). \tag{9.40}$$

For the example above, we get the following output for the training pattern:

$$\operatorname{sgn}\left(\frac{1}{3} \begin{bmatrix} 0 & -1 & -1 \\ -1 & 0 & -1 \\ -1 & -1 & 0 \end{bmatrix} \begin{bmatrix} -1 \\ 1 \\ -1 \end{bmatrix}\right) = \operatorname{sgn}\left(\begin{bmatrix} 0 \\ 2 \\ 0 \end{bmatrix}\right) = \begin{bmatrix} -1 \\ +1 \\ -1 \end{bmatrix}. \tag{9.41}$$

Therefore, the network evolves toward the desired attractor, and it is left as an exercise for the reader to show that the remaining patterns are also recovered.

The output of the Hopfield network is computed in Julia as:

```julia
for it in 1:M
    s = sgn.(W*s)
end
```

Hopfield networks can be used, for example, to fix corrupted images. See, for example, the patterns shown in the top row of Fig. 9.10 and consider an initial corrupted state corresponding to a representation of the number "2". The progression shown in the bottom row of the figure shows how the network can recover the correct pattern after a few steps.

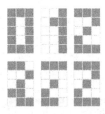

Figure 9.10 Top: patterns stored in the Hopfield network, Bottom: progression of recovery of a corrupted state

9.3.2 ENERGY AND ASYNCHRONOUS UPDATE

Drawing inspiration from physics models, specifically the *Ising model* [152], Hopfield attributed energy to this network (also known as the Lyapunov[p] function [153])

---

[p]Aleksandr Mikhailovich Lyapunov (1857–1918) Russian mathematician and physicist, advised by Pafnuty Chebyshev.

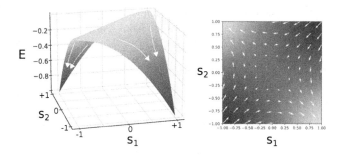

Figure 9.11 Energy profile for a two-node Hopfield network that remembers patterns $\mathbf{x}_1 = [-1, 1]$ and $\mathbf{x}_2 = [1, -1]$. The arrows show the flow toward the attractors

[147]. This energy can be found by multiplying the argument of the sign function in Eq. 9.40 by $\mathbf{s}^T$:

$$
\begin{aligned}
E &= -\frac{1}{2}\mathbf{s}^T\mathbf{W}\mathbf{s} + \mathbf{s}^T\mathbf{b} \\
&= -\frac{1}{2}\sum_{i,j}W_{ij}s_is_j + \sum_i b_is_i.
\end{aligned}
\tag{9.42}
$$

The energy landscape consists of several minima corresponding to the patterns that it can remember (*retrieval states*). Therefore, starting the network with values near one of these minima will make the network converge to the corresponding state.

Let's take, for instance, a two-node network that remembers the patterns $\mathbf{x}_1 = [-1, 1]$ and $\mathbf{x}_2 = [1, -1]$. The weight matrix is given by:

$$
\begin{aligned}
\mathbf{W} &= \frac{1}{2}\left(\begin{bmatrix} -1 \\ 1 \end{bmatrix}\begin{bmatrix} -1 & 1 \end{bmatrix} + \begin{bmatrix} 1 \\ -1 \end{bmatrix}\begin{bmatrix} 1 & -1 \end{bmatrix}\right) - \begin{bmatrix} 1 & 0 \\ 0 & 1 \end{bmatrix} \\
&= \frac{1}{2}\left(\begin{bmatrix} 1 & -1 \\ -1 & 1 \end{bmatrix} + \begin{bmatrix} 1 & -1 \\ -1 & 1 \end{bmatrix}\right) - \begin{bmatrix} 1 & 0 \\ 0 & 1 \end{bmatrix} \\
&= \begin{bmatrix} 0 & -1 \\ -1 & 0 \end{bmatrix}.
\end{aligned}
\tag{9.43}
$$

The energy associated with this weigh matrix is:

$$
\begin{aligned}
E &= -\frac{1}{2}\begin{bmatrix} s_1 & s_2 \end{bmatrix}\begin{bmatrix} 0 & -1 \\ -1 & 0 \end{bmatrix}\begin{bmatrix} s_1 \\ s_2 \end{bmatrix} \\
&= -\frac{1}{2}\begin{bmatrix} s_1 & s_2 \end{bmatrix}\begin{bmatrix} -s_2 \\ -s_1 \end{bmatrix} \\
&= s_1 s_2.
\end{aligned}
\tag{9.44}
$$

This energy landscape is shown in Fig. 9.11. Observe that the two patterns correspond to two minima of the energy landscape and any state is pushed toward those attractors as indicated by the arrows in the figure.

When updating the network, its energy can either stay unchanged or reduce. To prove the latter let's consider a change of state from s_j to s'_j:

$$\Delta_E = -\frac{1}{2}\sum_i W_{ij}s_is_j + \phi_is_j + \frac{1}{2}\sum_i W_{ij}s_is'_j - b_is_j$$

$$= -\frac{1}{2}(s_j - s'_j)\left(\sum_j w_{ij}s_i - b_i\right). \tag{9.45}$$

Since the state changed, s'_j has a sign opposite to that of s_j. Therefore, Δ_E is positive and the energy is reduced for every change until it reaches a minimum.

Given an energy function, it is possible to asynchronously update the network in a Monte Carlo fashion:

```
while(energy1 > threshold)
    p = rand(1:N)
    t = copy(s)
    t[p] = -t[p]
    Energy2 = (-0.5*t'*W*t)[1]

    if (Energy2 < Energy1)
        s = copy(t)
        Energy1 = Energy2
    end
end
```

9.3.3   CAPACITY AND CROSS-TALK

For a pattern $\mathbf{x}_j$, we get an output:

$$s^{l+1} = \text{sgn}\left[\left(\frac{1}{M}\sum_{i=1}^{M}\mathbf{x}_i\mathbf{x}_i^T - \mathbf{I}\right)\mathbf{x}_j\right]$$

$$= \text{sgn}\left[\frac{1}{M}\mathbf{x}_j\mathbf{x}_j^T\mathbf{x}_j - \mathbf{x}_j + \frac{1}{M}\sum_{i\neq j}^{M}\mathbf{x}_i\mathbf{x}_i^T\mathbf{x}_j\right] \tag{9.46}$$

$$= \text{sgn}\left[\frac{1}{M}\left((N-M)\mathbf{x}_j + \sum_{i\neq 0}\mathbf{x}_i(\mathbf{x}_i\cdot\mathbf{x}_j)\right)\right].$$

The first property we extract from this equation is that the number of nodes N in a Hopfield network has to be bigger than the number of patterns M to be stored. The second term inside the sign function is known as *cross-talk*. Hence, if all patterns are orthogonal, the attractors are stable. However, in the presence of a correlation between the patterns, the network becomes unstable.

Figure 9.12 shows the effect of cross-talk. The figure on the top shows an arbitrary energy profile for two stored patterns x_1 and x_2. It is perfectly possible to distinguish the two minima in this configuration. However, as the patterns are brought closer together, their energy profiles merge and it becomes difficult to distinguish between them. This situation can be overcome if a different energy function is used as shown in Fig. 9.12c.

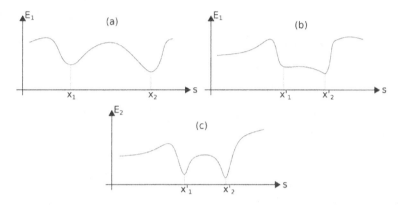

Figure 9.12 a) An arbitrary energy profile with two local minima corresponding to two stored patterns x_1 and x_2. b) If the two states are correlated, they may produce a cross-talk effect. c) A different energy function for the same stored patterns that mitigates cross-talking

The storage capacity of patterns free of errors in a Hopfield network is given by [154]:

$$C \approx \frac{N}{2\log(N)} \approx 0.14N. \tag{9.47}$$

Note that the example of Fig. 9.10 has 15 nodes. Therefore, fewer than three patterns can be correctly stored.

It is possible to increase the capacity of the network by modifying the energy in such a way that it reduces the correlation between the stored patterns:

$$E = -\sum_{i=1}^{M} f\left(\mathbf{x}_i^T \mathbf{s}\right), \tag{9.48}$$

where f is a smooth function. When the energy is modified this way, the network is known as a *dense associative memory* (DAM) [155, 156].

For instance, one can use:

$$E = -\sum_{i=1}^{M} \exp\left(\mathbf{x}_i^T \mathbf{s}\right). \tag{9.49}$$

If we construct a matrix $\mathbf{W} = [\mathbf{x}_1, \ldots, \mathbf{x}_M]^\top$, then this energy function can be written as[q]:

$$E = -\exp\left(lse(\mathbf{Ws}, 1)\right). \tag{9.50}$$

Another possibility considering continuous-valued patterns is to use an energy function described by [157]:

$$E = -lse\left(\mathbf{Ws}, \beta\right) + \frac{1}{2}\mathbf{s}^T\mathbf{s} + \frac{1}{\beta}\log(M) + \frac{1}{2}P^2, \tag{9.51}$$

where P is the largest norm of all memorized patterns $\mathbf{x}_i$. It is possible to obtain an update rule for this energy function by applying a concave-convex procedure if we notice that the term $\mathbf{s}^T\mathbf{s}$ is convex and the term $-lse\left(\mathbf{Ws}, \beta\right)$ is concave:

$$\frac{\partial}{\partial \mathbf{s}^{t+1}}\left(\frac{1}{2}\left(\mathbf{s}^{t+1}\right)^T \mathbf{s}^{t+1}\right) = -\frac{\partial}{\partial \mathbf{s}^t}\left(-lse\left(\mathbf{Ws}^t, \beta\right)\right)$$
$$\mathbf{s}^{t+1} = \mathbf{W}^\top \text{softmax}\left(\beta\mathbf{Ws}^t\right). \tag{9.52}$$

Under this description, $\mathbf{W}$ can be understood as weights between the state $\mathbf{s}$ and hidden units, which we will call $\mathbf{h}^t$. By the same token, $\mathbf{W}$ can be viewed as a matrix of weights between hidden units and the state $\mathbf{s}$ [158]. This is illustrated in Fig. 9.13.

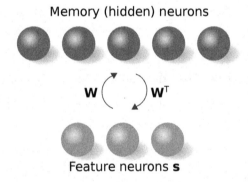

Memory (hidden) neurons

$\mathbf{W}$ $\mathbf{W}^\mathsf{T}$

Feature neurons $\mathbf{s}$

Figure 9.13   Dense associative memory as a two-layer network

---

Concave-Convex Procedure (CCP) [159]

Consider an energy function composed of concave and convex terms: $E(\mathbf{x}) = E_{vex}(\mathbf{x}) + E_{cav}(\mathbf{x})$. For any points x_i, $i = 1, 2, 3, 4$, given the convexity and concavity of the terms, we can write:

$$E_{vex}(\mathbf{x}_2) \geq E_{vex}(\mathbf{x}_1) + (\mathbf{x}_2 - \mathbf{x}_1)\nabla E_{vex}(\mathbf{x}_1)$$
$$E_{cav}(\mathbf{x}_4) \leq E_{cav}(\mathbf{x}_3) + (\mathbf{x}_4 - \mathbf{x}_3)\nabla E_{cav}(\mathbf{x}_3)$$

---

[q]The log-sum-exp function is described by $lse(\mathbf{x}, \beta) = \beta^{-1}\log\left(\sum_i \exp(\beta x_i)\right)$.

> **Concave-Convex Procedure (CCP) [159] (Continued)**
>
> Setting $\mathbf{x}_1 = \mathbf{x}_4 = \mathbf{x}^{t+1}$ and $\mathbf{x}_2 = \mathbf{x}_3 = \mathbf{x}^t$, we get:
>
> $$\frac{E_{cav}(\mathbf{x}^{t-1}) - E_{cav}(\mathbf{x}^t)}{\mathbf{x}^{t+1} - \mathbf{x}^t} \leq \nabla E_{cav}(\mathbf{x}^t)$$
>
> $$\frac{E_{vex}(\mathbf{x}^{t-1}) - E_{vex}(\mathbf{x}^t)}{\mathbf{x}^{t+1} - \mathbf{x}^t} \leq \nabla E_{vex}(\mathbf{x}^{t+1})$$
>
> Taking the limit when $\mathbf{x}^{t+1} - \mathbf{x}^t \to 0$ and adding both expressions, we immediately see that:
>
> $$\nabla E_{vex}(\mathbf{x}^{t+1}) = -\nabla E_{cav}(\mathbf{x}^t).$$

This way, it is possible to rewrite the previous equation more explicitly as:

$$\mathbf{h}^t = \beta \mathbf{W} \mathbf{s}^t$$
$$\mathbf{s}^{t+1} = \mathbf{W}^\top \text{softmax}(\mathbf{h}^t). \tag{9.53}$$

It is interesting to note how the output of a DAM resembles the attention mechanism shown in Sec. 9.2.1.

In Julia, the output of the dense associative memory can be obtained with:

```
for it in 1:5
    h = β*W*s
    s = transpose(W)*softmax(h)
end
```

Using DAM, we can use the same network but add more patterns as shown in Fig. 9.14. While the standard Hopfield network in the previous example could only remember less than three patterns, DAM can remember patterns from bigger training sets.

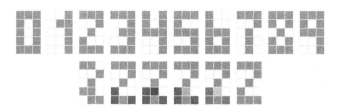

Figure 9.14 A bigger training set shown in the top row, and the recovery process of a corrupted pattern corresponding to "2" in the bottom row

9.4   BOLTZMANN MACHINES

A Boltzmann[r] machine [160] is a bidirectional fully connected network of artificial neurons with binary values $s_i \in \{0, 1\}$ as shown in Fig. 9.15. It is a stochastic version of the Hopfield network[s], and, consequently a Markov random field, where training follows a simulated annealing strategy (see Sec. 13.2.). In a *restricted* Boltzmann machine (RBM) [162] there are no hidden-hidden or visible-visible connections. Thus, the network of an RBM is a bipartite graph.

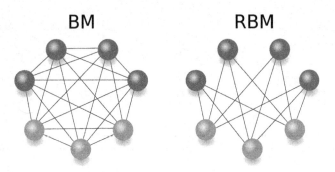

Figure 9.15   Left: Boltzmann machine (BM), Right: Restricted Boltzmann machine (RBM)

The training of restricted networks follows the principles of statistical mechanics. For example, the probability of finding the network in a state $\mathbf{x}, \mathbf{h}$ is given by:

$$P(\mathbf{x}, \mathbf{h}) = \frac{1}{Z} e^{-\varepsilon(\mathbf{x}, \mathbf{h})}, \tag{9.54}$$

where Z is the partition function given by:

$$Z = \sum_{\mathbf{x}, \mathbf{h}} e^{-\varepsilon(\mathbf{x}, \mathbf{h})} \tag{9.55}$$

and ε is an energy function given by:

$$\varepsilon(\mathbf{x}, \mathbf{h}) = -\mathbf{b} \cdot \mathbf{x} - \mathbf{c} \cdot \mathbf{h} - \mathbf{h} \cdot (\mathbf{W}\mathbf{x}). \tag{9.56}$$

In the last expression, $\mathbf{b}$ and $\mathbf{c}$ are the biases of the visible and hidden layers, respectively, whereas $\mathbf{W}$ is a connection matrix between the layers.

Just like in a Hopfield network, we want the Boltzmann network to remember a pattern in the visible layer. Therefore, we are more interested in the probability the

---

[r]Ludwig Eduard Boltzmann (1844–1906) Austrian physicist, advised by Josef Stefan, Gustav Kirchhoff, and Hermann von Helmholtz, among others. Boltzmann was the adviser of Paul Ehrenfest, among others.

[s]It is also a stochastic version of the Ising model, also known as the Sherrington[t]-Kirkpatrick[u] spin-glass model [161].

[t]David Sherrington (1941–Present) British physicist.

[u]Scott Kirkpatrick, American physicist.

network shows a certain pattern $\mathbf{x}$ in its input. For this, we marginalize the probability[v]:

$$
\begin{aligned}
P(\mathbf{x}) &= \sum_{\mathbf{h}} P(\mathbf{x}, \mathbf{h}) = \frac{1}{Z} \sum_{\mathbf{h}} e^{\mathbf{b} \cdot \mathbf{x} + \mathbf{c} \cdot \mathbf{h} + \mathbf{h} \cdot (\mathbf{Wx})}. \\
&= \frac{1}{Z} e^{\mathbf{b} \cdot \mathbf{x}} \sum_{\mathbf{h}} \prod_{i} \exp\left\{ [c_i + (\mathbf{Wx})_i] \, h_i \right\} \\
&= \frac{1}{Z} e^{\mathbf{b} \cdot \mathbf{x}} \prod_{i} \left(1 + e^{c_i + (\mathbf{Wx})_i} \right) \\
&= \frac{1}{Z} e^{-\varepsilon(\mathbf{x})},
\end{aligned}
\tag{9.57}
$$

where $\varepsilon(\mathbf{x})$ is an effective energy given by:

$$
\begin{aligned}
\varepsilon(\mathbf{x}) &= -\mathbf{b} \cdot \mathbf{x} - \sum_{i} \log\left(1 + e^{c_i + (\mathbf{Wx})_i} \right) \\
&= -\mathbf{b} \cdot \mathbf{x} - \sum_{i} \text{softplus}\,(c_i + (\mathbf{Wx})_i),
\end{aligned}
\tag{9.58}
$$

and *softplus* is an activation function. Also, since the partition function is a normalizing term, we can write $Z = \sum_{\mathbf{x}} e^{-\varepsilon(\mathbf{x})}$.

Now, the likelihood for the ith training data set $\mathbf{s}_i$ is:

$$
\mathscr{L}(\mathbf{s}_i) = P(\mathbf{x} = \mathbf{s}_i). \tag{9.59}
$$

To estimate parameters $\theta = \{\mathbf{b}, \mathbf{c}, \mathbf{W}\}$, we can maximize the log-likelihood[w]:

$$
\begin{aligned}
\log(\mathscr{L}(\mathbf{s}_i)) &= -\log(Z) - \varepsilon(\mathbf{x})|_{\mathbf{x} = \mathbf{s}_i} \\
\frac{\partial \log(\mathscr{L}(\mathbf{s}_i))}{\partial \theta} &= -\frac{1}{Z} \frac{\partial Z}{\partial \theta} - \frac{\partial \varepsilon(\mathbf{x})}{\partial \theta}\bigg|_{\mathbf{x} = \mathbf{s}_i} \\
&= \frac{1}{Z} \sum_{\mathbf{x}} e^{-\varepsilon(\mathbf{x})} \frac{\partial \varepsilon(\mathbf{x})}{\partial \theta} - \frac{\partial \varepsilon(\mathbf{x})}{\partial \theta}\bigg|_{\mathbf{x} = \mathbf{s}_i} \\
&= \sum_{\mathbf{x}} P(\mathbf{x}) \frac{\partial \varepsilon(\mathbf{x})}{\partial \theta} - \frac{\partial \varepsilon(\mathbf{x})}{\partial \theta}\bigg|_{\mathbf{x} = \mathbf{s}_i}.
\end{aligned}
\tag{9.60}
$$

Approximating the expected value of the target distribution by the average over the whole training set $\mathbf{s}$:

$$
\begin{aligned}
\left\langle \frac{\partial \log(\mathscr{L}(\mathbf{s}_i))}{\partial \theta} \right\rangle_{\mathbf{s}} &= \left\langle \frac{\partial \varepsilon(\mathbf{x})}{\partial \theta} \right\rangle_{\mathbf{x}} - \frac{1}{N} \sum_{j=1}^{N} \frac{\partial \varepsilon(\mathbf{x})}{\partial \theta}\bigg|_{\mathbf{x} = \mathbf{s}_j} \\
&= \left\langle \frac{\partial \varepsilon(\mathbf{x})}{\partial \theta} \right\rangle_{\mathbf{x}} - \left\langle \frac{\partial \varepsilon(\mathbf{x})}{\partial \theta} \right\rangle_{\mathbf{s}}.
\end{aligned}
\tag{9.61}
$$

---

[v] Here we use: $a_1 b_1 + a_1 b_2 + a_2 b_1 + a_2 b_2 = (a_1 + a_2)(b_1 + b_2) \rightarrow \sum_{\mathbf{h}} \prod_i f(h_i) = \prod_i \sum_{h_i} f(h_i)$.
[w] This is equivalent to minimizing the Kullback-Leibler divergence, see Sec. 2.2.1.

The gradients computed for the training set are simple to find:

$$\frac{\partial \varepsilon(\mathbf{x})}{\partial \mathbf{b}}\bigg|_{\mathbf{x}=\mathbf{s}_j} = -\mathbf{s}_j^\top$$

$$\frac{\partial \varepsilon(\mathbf{x})}{\partial c_i}\bigg|_{\mathbf{x}=\mathbf{s}_j} = -\sigma\left(c_i + (\mathbf{W}\mathbf{s}_j)_i\right) \rightarrow \frac{\partial \varepsilon(\mathbf{x})}{\partial \mathbf{c}}\bigg|_{\mathbf{x}=\mathbf{s}_j} = -\mathbf{y}^\top \qquad (9.62)$$

$$\frac{\partial \varepsilon(\mathbf{x})}{\partial W_{ij}}\bigg|_{\mathbf{x}=\mathbf{s}_j} = -(\mathbf{s}_j)_i\sigma\left(c_i + (\mathbf{W}\mathbf{s}_j)_i\right) \rightarrow \frac{\partial \varepsilon(\mathbf{x})}{\partial \mathbf{W}}\bigg|_{\mathbf{x}=\mathbf{s}_j} = -\mathbf{s}_j\mathbf{y}^\top,$$

where $\sigma(x) = (1 + e^{-x})^{-1}$ is the sigmoid activation.

In Julia:

```julia
y0 = σ.(c + W*s)
y1 = σ.(c + W*x)

∇b = s - x
∇c = y0 - y1
∇W = s*transpose(y0) - x*transpose(y1)
```

The problem is how to compute the terms that involve the partition function and the sum over all possible $\mathbf{h}$ and $\mathbf{x}$. One idea is to approximate the expectation by an estimator using Gibbs[x] sampling [163]. We begin by randomly choosing M vectors from the training set and then initializing the visible layer with one of them. Next, we set the states of the hidden layer to 1 with probability $P(h_i = 1|\mathbf{x})$ and reconstruct the visible layer by setting its states to 1 with probability $P(x_i = 1|\mathbf{h})$. Once all samples were computed, we can estimate the parameters θ. The procedure is finally repeated until convergence has been reached.

The conditional probabilities are computed as:

$$P(\mathbf{h}|\mathbf{x}) = \frac{P(\mathbf{x},\mathbf{h})}{P(\mathbf{x})} = \frac{\prod_j e^{[c_j + (\mathbf{W}\mathbf{x})_j]h_j}}{\prod_{j'}\left(1 + e^{c_{j'} + (\mathbf{W}\mathbf{x})_{j'}}\right)}$$

$$= \prod_j \frac{e^{[c_j + (\mathbf{W}\mathbf{x})_j]h_j}}{1 + e^{c_j + (\mathbf{W}\mathbf{x})_j}} = \prod_j P(h_j|\mathbf{x}) \qquad (9.63)$$

$$\therefore P(h_j = 1|\mathbf{x}) = \sigma\left(c_j + (\mathbf{W}\mathbf{x})_j\right).$$

Similarly,

$$P(x_j = 1|\mathbf{h}) = \sigma\left(b_j + (\mathbf{h}^\top\mathbf{W})_j\right). \qquad (9.64)$$

---

[x]Josiah Willard Gibbs (1839–1903) American polymath winner of many prizes, including the Copley Medal in 1901. Gibbs advised many students, including Lee de Forest and Irving Fisher.

This not only confirms that the hidden variables are conditionally independent given any visible layer but it also makes an interesting parallel with a neural network. This in Julia can be computed as:

```
for i in 1:N
    p = σ(c[i] + W[i,:]·s)
    if (rand() < p)
        h[i] = 1
    else
        h[i] = 0
    end
end
```

With this, we arrive at the *contrastive divergence* algorithm [162] below:

Algorithm 18: Contrastive Divergence (CD-k)

---

repeat
   ▷ Select k vectors from the training dataset (a mini-batch);
   ▷ Initialize the visible states with the training data;
   for k *times* **do**
       ▷ Calculate the hidden states setting them to 1 with the probability given by Eq. 9.63;
       ▷ Reconstruct the input states with probability given by Eq. 9.64;
   end
   ▷ Use the reconstruction to estimate $\nabla_\theta \varepsilon$ and update θ according to Eq. 9.61;
until *convergence*;

Section V

Reinforcement Learning

10 Value Methods

During the 1890s, Pavlov[a] conducted an interesting study about conditioning [164]. In his study, Pavlov gave food to dogs and observed their salivation. To his surprise, he found that the dogs developed salivation every time they heard the footsteps of the lab assistant bringing the food. This developed conditioning is a classic example of reinforcement learning where the dog learned an association between footsteps and food. The food behaves as a reward that was paired with the footsteps being a stimulus.

In general, the idea of reinforcement learning [165] is of an agent that takes an action a_t upon an environment and the result of this action is fed back to the agent via a reward r_t and a description of the current state s_t. The rule that the agent uses to select actions is called *behavior*, which creates a state, an action, and a reward sequence $\{(s_t, a_t, r_{t+1}), t \geq 0\}$. This general idea is shown in Fig. 10.1.

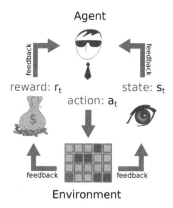

Figure 10.1 General structure of a reinforcement learning algorithm

Reinforcement learning encompasses several concepts that can be highly abstract. To make it easier to understand these concepts, let's consider a simple game throughout this chapter. The game, as illustrated in Fig. 10.2, consists of a prize hidden in a maze, and the player must find it. At any instant, the agent is only aware of the current state and must make a decision about what action to take, which can be to move up, down, left, or right until the prize is found. This is captured by the

[a] Ivan Petrovich Pavlov (1849–1936) Russian physiologist, winner of the Nobel Prize in Physiology or Medicine in 1904.

DOI: 10.1201/9781003350101-10

following matrix:

$$M = \begin{bmatrix} \times & \times & \times & B \\ \times & D & A & \times \\ \times & \times & \times & D \\ B & \times & C & \times \end{bmatrix} \tag{10.1}$$

where the rows indicate the state, the columns indicate the action, and $\times$ indicates that that transition will never occur. So for example, if the agent is in state A, the only possible move is to move right and arrive in state B. Also, in this game, C is an absorbing state.

Also, depending on the state the agent arrives, a different reward is obtained.

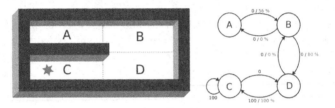

Figure 10.2   A simple 2×2 maze for a game where the player has to find the prize located in C (left) and the corresponding Markov chain (right). The numbers in the arrows are the immediate reward and the q-value, respectively

One framework the agent can use for taking decisions is the *Markov Decision Process* (MDP) [166]. An MDP is given by a quadruple $(S, A, P, \mathscr{R})$, where S is a countable non-empty set of states (*state space*), and A is a countable non-empty set of actions (*action space*). In our example, $S = \{A, B, C, D\}$ and[b] $A = \{$move up, move down, move left, move right$\}$. Also, $P : S \times S \times A \to [0, 1]$ is a *transition probability function* that obeys the Markov property $P(s_{t+1}|s_0, \ldots, s_t, a_0, \ldots, a_t) = P(s_{t+1}|s_t, a_t)$ (see Sec. 2.4) that gives the probability of transitioning to state $s_{t+1} \in S$ departing from state $s_t \in S$ given an action $a \in A$. Our objective is to find this probability.

In the MDP, $\mathscr{R} : S \times A \to \mathbb{R}$ is an *immediate reward* received by the agent when taking action $a \in A$ when in state $s_t \in S$. The MDP may have absorbing (or terminal) states $s_0 \in S$ such that $s_{t'} = s_0|s_t = s_0, \forall t' \geq t$ (state C in the example). MDPs that contain such states are called *episodic* and an *episode* corresponds to reaching an absorbing state when departing from an initial state. The reward for our example can be described by the matrix:

$$R = \begin{bmatrix} -1 & -1 & -1 & 0 \\ -1 & 0 & 0 & -1 \\ -1 & -1 & -1 & 0 \\ 0 & -1 & 100 & -1 \end{bmatrix}, \tag{10.2}$$

where the rows describe the current state and the columns describe the action. The decision to take a prohibited action implies a negative reward, whereas taking an

---

[b]Care must be taken since A can represent either a state or the action space.

allowed action but landing in a state without a prize implies a zero reward. Finally, when the agent is on state D and moves left, he is rewarded with the prize $r = 100$.

The policy is a function $\pi : S, A \rightarrow \mathbb{R}$ that gives the probability of taking action a given state s, $\pi(a|s) = P(A_t = a|S_t = s)$. The objective is to find a behavior $\pi(a_t|s_t)$ that specifies a sequence of actions that maximizes the cumulative *reward*. For an *episodic task* this is given by:

$$R = r_1 + r_2 + \ldots + r_T, \tag{10.3}$$

where T is the period of an episode. We can define the cumulative reward for a specific time as:

$$R_t = r_{t+1} + r_{t+2} + \ldots + r_T = \sum_{k=1}^{T-t} r_{t+k}. \tag{10.4}$$

States that maximize the cumulative reward are said to be *optimal*. For example, if the game begins with the agent in state A, the optimal sequence would be to move right, down, and then left. This would imply a cumulative reward of $R_3 = 0 + 0 + 100$.

The agent may have a temporal preference toward present rewards and we may have to discount future rewards. Therefore, we rewrite the *discounted reward* as:

$$R_t = r_{t+1} + \gamma r_{t+2} + \gamma^2 r_{t+3} + \ldots + \gamma^{T-1} r_T = \sum_{k=1}^{T-t} \gamma^{k-1} r_{t+k}.$$
$$= r_{t+1} + \gamma R_{t+1}. \tag{10.5}$$

where $\gamma \in [0,1]$ is a *discount factor* that controls the agent's time preference. For instance, if $\gamma = 0$, then the agent only considers the immediate reward, whereas if $\gamma = 1$, then the agent weights equally the present and all future rewards.

A *stationary policy* is a mapping $\pi : S \rightarrow A$ that maps states to actions. This can be deterministic ($a_t = \pi(s_t)$) of stochastic ($\pi(a|s)$). The latter produces a time-homogeneous Markov chain $\{s_t, t \geq 0\}$ and the collection of all stationary policies is denoted by Π_{stat}.

To find the optimal behavior, one could consider testing all possible behaviors and then choosing one that gives the highest reward. This, however, is not feasible because of the big number of possible behaviors. Instead, it is possible to find the optimal behavior by mapping each state/action pair to a *state-value function* $V : S \rightarrow \mathbb{R}$ that gives the expected return when departing from state s and following a policy π:

$$V(s) = \langle R_t | s_t = s \rangle, \; s \in S. \tag{10.6}$$

It is also useful to define an *action-value function* $Q : S \times A \rightarrow \mathbb{R}$ that gives the expected return when departing from state s, following a policy π, and taking action a:

$$Q(s,a) = \langle R_t | s_t = s, a_t = a \rangle, \; s \in S, a \in A. \tag{10.7}$$

Each of these functions leads to different reinforcement learning strategies, but before discussing them, we must study an optimization strategy that can be used to make informed decisions, namely dynamic programming [167]. This, however, requires the use of Bellman[c]'s equations [168] to relate the state-value and the action-value functions.

10.1   DYNAMIC PROGRAMMING

The state-value function is related to the action-value function through:

$$
\begin{aligned}
V(s,t) &= \langle R_t | s_t = s \rangle \\
&= \sum_{r \in \mathscr{R}} p(r | s_t = t) r \\
&= \sum_{r \in \mathscr{R}} \sum_{a \in \mathscr{A}} p(r,a | s_t = s) r, \text{ conditioning:} \\
&= \sum_{a \in \mathscr{A}} p(a | s_t = s) \sum_{r \in \mathscr{R}} p(r | s_t = s, a_t = a) r \\
&= \sum_{a \in \mathscr{A}} p(a | s_t = s) \langle R_t | s_t = s, a_t = a \rangle \\
&= \sum_{a \in \mathscr{A}} p(a | s_t = s) Q(s,a).
\end{aligned}
\tag{10.8}
$$

The opposite relationship is given by:

$$
\begin{aligned}
Q(s,a) &= \langle R_t | s_t = s, a_t = a \rangle \\
&= \langle r_t + \gamma R_{t+1} | s_t = s, a_t = a \rangle, \text{ total expectation:} \\
&= \sum_{s' \in \mathscr{S}} \langle r_t + \gamma R_{t+1} | s_{t+1} = s', s_t = s, a_t = a \rangle p\left(s_{t+1} = s' | s_t = s, a_t = a \right) \\
&= \sum_{s' \in \mathscr{S}} p(s' | s,a) \left[r(s,a,s') + \gamma V(s',t+1) \right].
\end{aligned}
\tag{10.9}
$$

The last equation can conveniently be written as:

$$
Q^{\pi}(s,a) = \sum_{s' \in \mathscr{S}} p(s' | s,a) \left[r(s,a,s') + \gamma \sum_{a' \in \mathscr{A}} \pi(a' | s) Q^{\pi}(s',a') \right].
\tag{10.10}
$$

Also, Eqs. 10.8 and 10.9 can be combined into:

$$
V^{\pi}(s,t) = \sum_{a \in \mathscr{A}} \pi(a | s) \sum_{s' \in \mathscr{S}} p(s' | s,a) \left[r(s,a,s') + \gamma V^{\pi}(s',t+1) \right].
\tag{10.11}
$$

These last two equations are known as *Bellman's equations* [167, 168] and are illustrated by the *backup diagrams* [169] in Fig. 10.3. There are also Bellman's equations in continuous-time, which have implications to important problems in economics (see Appendix G).

We are interested in finding a policy corresponding to a set of actions taken by the agents that produces the highest reward:

$$
V^*(s) = \max_{\pi} V^{\pi}(s).
\tag{10.12}
$$

---

[c]Richard Ernest Bellman (1920–1984) American mathematician winner of many prizes, including the John von Neumann Theory Prize in 1976.

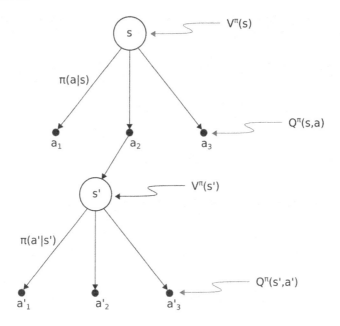

Figure 10.3   Backup diagram for Bellman's equations

This policy is known as an *optimal policy*. Alternatively, this is also given by the expected return for the best action. According to Eq. 10.9:

$$
\begin{aligned}
V^*(s) &= \max_{a \in \mathscr{A}} Q^\pi(s,a) \\
&= \max_{a \in \mathscr{A}} \sum_{s' \in \mathscr{S}} p(s'|s,a)\left[r(s,a,s') + \gamma V^*(s')\right].
\end{aligned}
\tag{10.13}
$$

Conversely, for the action-value function we have, according to Eq. 10.10:

$$
\begin{aligned}
Q^*(s,a) &= \max_\pi Q^\pi(s,a) \\
&= \sum_{s' \in \mathscr{S}} p(s'|s,a)\left[r(s,a,s') + \gamma \max_{a'} Q^*(s',a')\right].
\end{aligned}
\tag{10.14}
$$

From this, we can extract a *greedy policy*[d] [170], a probability of taking action a when in the state s:

$$
\pi(s,a) = \begin{cases} 1, & \text{if } a = \arg\max_a Q(s,a) \\ 0, & \text{otherwise.} \end{cases}
\tag{10.15}
$$

Given the recursive nature of Bellman's equations, we can use dynamic programming to evaluate a policy and then improve it.

---

[d]A greedy strategy chooses the immediate optimal choice at each step.

Dynamic Programming

Given any natural number n, in how many D_n ways can we write n as sums of the numbers 1, 2, and 3? The simplest approach to solve this problem is by brute force writing down all possibilities. For instance, for $n = 4$ we have:

$$
\begin{aligned}
&1+1+1+1 = 4, \quad 2+1+1 = 4, \\
&1+1+2 = 4, \quad\;\;\; 2+2 = 4, \\
&1+2+1 = 4, \quad\;\;\; 3+1 = 4, \\
&1+3 = 4.
\end{aligned}
\tag{10.16}
$$

Therefore, $D_4 = 7$. A simpler way to solve this problem would be to find a recurrence relation. For example, we can write $x_1 + x_2 + \ldots + x_N = n$. But if we know that, and we set $x_1 = 1$, then $x_2 + \ldots + x_N = n - 1$. Therefore, there must be D_{n-1} ways to write $n - 1$ as sums of 1, 2, and 3. We can do the same procedure setting $x_1 = 2$ and find D_{n-2} ways to write $n - 2$ as sums of 1, 2, and 3. Finally, we do the same for $x_1 = 3$ and conclude that $D_n = D_{n-1} + D_{n-2} + D_{n-3}$. Therefore:

$$
\begin{aligned}
D_1 &= D_0 = 1 \\
D_2 &= D_1 + D_0 = 1 + 1 = 2 \\
D_3 &= D_2 + D_1 + D_0 = 2 + 1 + 1 = 4 \\
D_4 &= D_3 + D_2 + D_1 = 4 + 2 + 1 = 7.
\end{aligned}
\tag{10.17}
$$

This way of breaking down a problem in sub-problems that can be solved recursively is known as *dynamic programming* [167].

Given a policy π, we want to estimate the state-value and the action-value functions. According to Eq. 10.11, the state-value function can be rewritten as:

$$
V_{n+1}(s) = \sum_{a \in \mathcal{A}} \pi(a|s) \sum_{s' \in \mathcal{S}} p(s'|s,a) \left[r(s,a,s') + \gamma V_n(s') \right],
\tag{10.18}
$$

with V^π being a fixed point of the interaction. On the other hand, we do not know $p(s'|s,a)$ yet.

10.2   MONTE CARLO AND TEMPORAL DIFFERENCE

How can we estimate the value functions if we don't have a model? One possibility is, given Eq. 10.6:

$$
V(s) = \langle R_t | s_t = s \rangle, \; s \in \mathcal{S},
\tag{10.19}
$$

we obtain an empirical average by running N simulations departing from any random state s and running the same policy π:

$$
\begin{aligned}
\hat{V}_N(s) &= \frac{1}{N}\sum_{n=1}^{N} R^n(s) = \frac{1}{N}\left(\sum_{n=1}^{N-1} R^n(s) + R^N(s)\right) \\
&= \frac{N-1}{N}\frac{1}{N-1}\sum_{n=1}^{N-1} R^n(s) + \frac{1}{N}R^N(s) \\
&= \left(1 - \frac{1}{N}\right)\hat{V}_{N-1}(s) + \frac{1}{N}R^N(s) \\
&= \hat{V}_{N-1}(s) + \frac{1}{N}\left(R^N(s) - \hat{V}_{N-1}(s)\right).
\end{aligned}
\tag{10.20}
$$

In this equation, the future reward is given by Eq. 10.5. By taking the limit $N \to \infty$, the empirical average should converge to V^π.

Although it is possible to obtain many $\hat{V}_N(s)$ from a single run, the random $R^N(s)$ variables may be correlated. Another problem with this approach is that we need to reach the end of an episode to estimate the value function. To avoid this, let's work with Eqs. 10.5 and 10.6 a little bit more:

$$
\begin{aligned}
V(s_t) &= \langle R_t | s_t = s \rangle \\
&= \langle r_{t+1} + \gamma R_{t+1} | s_t = s \rangle \\
&= r_{t+1} + \gamma\langle R_{t+1} | s_t = s \rangle \\
&\approx r_{t+1} + \gamma V(s_{t+1}).
\end{aligned}
\tag{10.21}
$$

Therefore, we can rewrite the Monte Carlo estimate (Eq. 10.20) as:

$$
\begin{aligned}
\hat{V}_n(s_t) &\approx \hat{V}_{n-1}(s_t) + \alpha\left(r_{t+1} + \gamma\hat{V}_{n-1}(s_{t+1}) - \hat{V}_{n-1}(s)\right) \\
&\approx (1-\alpha)\hat{V}_{n-1}(s_t) + \alpha\left[r_{t+1} + \gamma\hat{V}_{n-1}(s_{t+1})\right].
\end{aligned}
\tag{10.22}
$$

The same can be applied to the action-value function:

$$
\hat{Q}_n(s_t, a_t) \approx (1-\alpha)\hat{Q}_{n-1}(s_t, a_t) + \alpha\left[r_{t+1} + \gamma\hat{Q}_{n-1}(s_{t+1}, a_{t+1})\right].
\tag{10.23}
$$

This gives a *temporal difference* (TD) [171] error $\delta = r_{t+1} + \gamma Q_{n+1}(s_{t+1})$ that describes the current surprise. It can be calculated immediately without having to wait until the end of an episode. For instance, $\delta > 0$ implies that the transition was positively surprising leading to a better reward than expected.

When improving the policy, it is possible to either use the improvement to generate the next episode or generate the next episode based on a different policy. Whereas the former is known as *on-policy*, the latter is known as *off-policy*. On-policy TD control is also known as state-action-reward-state-action (SARSA) (can be read off straight from the right-hand side of Eq. 10.23), whereas off-policy TD control is also known as Q-learning [169]. Before exploring both strategies, we need to deal with the problem of exploitation *vs.* exploration.

10.2.1 EXPLOITATION *VS.* EXPLORATION

At the beginning of a learning process, the agent does not know the environment and needs to explore it. If only the greedy solution is taken (exploitation), there is no exploration and the agent might get stuck in a local minimum. When exploiting a solution, the agent maximizes its immediate reward, but when exploring, the agent gets better rewards in the long run. Therefore, instead of selecting the highest esti-mate reward all the time ($\max_a Q$), the agent can choose a random action with a small probability ε.

The balance between exploration and exploitation is illustrated by the *multi-armed bandit* problem [172]. Imagine a situation where you are faced with different slot machines that have different success probabilities $\{\theta_1, \ldots, \theta_N\}$ but you do not know their behaviors before playing. Given limited resources, how do you distribute them among the different slot machines to obtain the highest reward? You begin playing in a machine and receive a reward r. You have a set of possible actions A that you can take on the ith slot machine. The value of each action is the expected reward $Q(a) = \langle r|a \rangle = \theta_i$. Since we are considering a slot machine, the reward is given by $r_t = R(a_t)$, where R is a reward function. It gives "1" with probability $Q(a_t)$ or "0" otherwise. This constitutes a Bernoulli process described by the ordered pair $(\mathscr{A}, R)$, which can also be considered to be a simplified MDP without a state space.

The goal, as usual, is to maximize the cumulative reward using an optimal policy. But how can we find it? One possibility is to take random actions and try to figure out the reward probability of each machine. This exploration, however, takes time and is likely to incur some losses. It is also possible to exploit the best action so far, but there is no guarantee that this will always be the best action.

The ε-greedy algorithm balances both strategies by selecting a random action with a small probability ε but also selecting the action with the highest estimated reward most of the time with probability $1 - \varepsilon$.

A strategy is said to be ε-*soft* if each action is taken with probability $\pi(a|s) \geq \varepsilon/\|\mathscr{A}(s)\|$ [169]. In the ε-greedy algorithm, there are $\|\mathscr{A}(s)\| - 1$ nongreedy ac-tions to be taken. Therefore, the probability of picking a nongreedy action is $\varepsilon(\|\mathscr{A}(s)\| - 1)/\|\mathscr{A}(s)\|$. Consequently, the greedy action is taken with probability:

$$\pi(a|s) = 1 - \varepsilon \frac{\|\mathscr{A}(s)\| - 1}{\|\mathscr{A}(s)\|} = 1 - \varepsilon + \frac{\varepsilon}{\|\mathscr{A}(s)\|}. \tag{10.24}$$

We then take the greedy action with probability $1 - \varepsilon$ but each of the remaining $|\mathscr{A}(s)|$ strategies will be taken with equal probability $\varepsilon/|\mathscr{A}(s)|$.

Therefore, to select an action, the ε-greedy algorithm reads [169, 170, 173]:

Algorithm 19: ε-greedy Action Selection

Input: Q table
Function *SelectAction(state,ε)* **is**
    $\triangleright$ r $\leftarrow$ uniform random number between 0 and 1;
    if $r < \varepsilon$ **then**
        ! Exploration
        $\triangleright$ action $\leftarrow$ valid random action from action space;
    else
        ! Exploitation
        $\triangleright$ action $\leftarrow \arg\max_a Q(state,a)$;
    end
end
Output: action

This algorithm is very simple to implement in Julia:

```julia
if (r < ε)
    a = rand(1:size(Q,2))
else
    a = argmax(Q[s,:])
end
```

Another possibility is to use the softmax function distribution as the probability of choosing an action a given state s:

$$p(a|s) = \frac{\exp\{\beta Q(s,a)\}}{\sum_{n\in\mathscr{A}}\exp\{\beta Q(s,n)\}}, \tag{10.25}$$

where β is the inverse temperature. Small values of β lead to a more homogeneous distribution, whereas big values lead to more concentrated distributions.

10.3 SARSA AND Q-LEARNING

Given the ε-greedy algorithm and the time difference equations, we can devise on-policy and off-policy TD control algorithms.

The first, SARSA [169] is an on-policy algorithm that can be derived straight from Eq. 10.23. There, the next episode is generated directly from the improved policy:

Algorithm 20: SARSA

  Input: state
  for *each episode* **do**
    ▷ Choose state s, and action a derived from the ε-greedy algorithm;
    repeat
      ▷ Take action a and obtain reward r and next state s';
      ▷ Choose action a' derived from the ε-greedy algorithm;
      ▷ $Q(s,a) \leftarrow (1-\alpha)Q(s,a) + \alpha\left[r + \gamma Q(s',a')\right]$;
      ▷ Update: $s \leftarrow s'$ and $a \leftarrow a'$;
    until *a terminal state is reached*;
  end

The main loop in Julia becomes:

```
s = rand(1:number_of_states)
a = ε_greedy(s,0.1,Q)

while(s ≠ absorbing_state)
    r = reward[s,a]
    sn = next_state[s,a]
    an = ε_greedy(s,0.1,Q)

    Q[s,a] = (1-α)*Q[s,a] + α*(r+γ*Q[sn,an])
    s = sn
    a = an
end
```

The Bellman equation for the action-value function (Eq. 10.14) is:

$$Q^*(s,a) = \sum_{s' \in \mathscr{S}} p(s'|s,a)\left[r(s,a,s') + \gamma \max_{a'} Q^*(s',a')\right].$$

It is possible to adapt Eq. 10.23 to obtain the same strategy of using the greedy action to update the current transition. Hence, it learns an optimal policy independent of the policy the agent is following, and, consequently, it is considered an off-policy strategy. This is the essence of the Q-learning algorithm [169]:

Algorithm 21: Q-learning
Input: state

for *each episode* **do**
 ▷ Choose a state s;
 repeat
  ▷ Choose an action a derived from the ε-greedy algorithm;
  ▷ Take action a and obtain reward r and next state s';
  ▷ $Q(s,a) \leftarrow (1-\alpha)Q(s,a) + \alpha \left[r + \gamma \max_{a' \in \mathscr{A}} Q(s',a') \right]$;
  ▷ Update: $s \leftarrow s'$;
 until *a terminal state is reached*;
end

In Julia, Q-learning becomes:

```
s = rand(1:number_of_states)

while(s ≠ absorbing_state)
    a = ε_greedy(s,0.1,Q)
    r = reward[s,a]
    sn = next_state[s,a]

    Q[s,a] = (1-α)*Q[s,a] + α*(r+γ*maximum(Q[sn,:]))
    s = sn
end
```

Both algorithms produce very close results for the simple game in our example. For example, after 500 episodes using $\alpha = 0.7$ and $\gamma = 0.9$, the Q matrix obtained with SARSA is:

$$Q \approx \begin{bmatrix} 53.73 & 63.26 & 71.18 & 79.86 \\ 63.60 & 89.67 & 71.22 & 84.17 \\ 0.72 & 0.17 & 0.42 & 0.32 \\ 53.88 & 94.38 & 100.40 & 94.37 \end{bmatrix}. \tag{10.26}$$

From this, the agent would preferably move right from state A, down from state B, and left from state D, which is expected.

One inconvenience of value-based approaches is that one needs a value matrix, that depending on the problem, can be prohibitively large. This can be circumvented by estimating the value function with a neural network. To use it, we need the TD error to approach zero. This can be achieved with the following losses for SARSA and Q-learning, respectively:

$$\mathscr{L}_{SARSA}(\theta) = \left\langle \left(r + \gamma\, Q_\theta(s',a') - Q_\theta(s,a) \right)^2 \right\rangle$$
$$\mathscr{L}_Q(\theta) = \left\langle \left(r + \gamma \max_{a' \in \mathscr{A}} Q_\theta(s',a') - Q_\theta(s,a) \right)^2 \right\rangle \tag{10.27}$$

When using a neural network to estimate the action-value function, it is possible to either design a network that takes both the state and action as inputs and returns the Q value, or one that returns all values of Q for every single (finite) action. These two strategies are illustrated in Fig. 10.4.

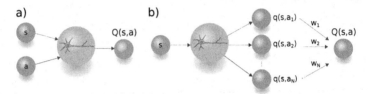

Figure 10.4 Action-value function approximations: a) One Q as a function of state and specific action, and b) One Q for each possible action for a single input state

When using a network that returns all values $Q(s,a)$, one must pay attention to the backpropagation rules. In the forward state, we have:

$$Q_\theta(s,a) = \mathbf{w} \cdot \mathbf{q}_\theta(s,\mathbf{a}), \tag{10.28}$$

where $\mathbf{w}$ is a vector that selects the ath element from $\mathbf{q}_\theta(s,\mathbf{a})$. Therefore, the backpropagation must be:

$$\frac{\partial \mathscr{L}}{\partial \theta} = \frac{\partial L}{\partial Q(s,a)} \frac{\partial Q(s,a)}{\partial \mathbf{q}_\theta(s,\mathbf{a})} \frac{\partial \mathbf{q}(s,\mathbf{a})}{\partial \theta}$$
$$= 2\left(Q(s,a) - y \right) \mathbf{w} \cdot \frac{\partial \mathbf{q}(s,\mathbf{a})}{\partial \theta}, \tag{10.29}$$

where $y = r + \gamma \max_{a' \in \mathscr{A}} Q_\theta(s',a')$ for Q-learning. This means that only the output $q(s,a)$ receives the backpropagation.

Also, when using neural networks to minimize those costs, one must pay attention to two problems. First, steps of the same episode can be correlated and this poses a big problem for training a neural network. Typically, this can be solved by using a huge (*replay*) buffer and sampling mini-batches from it. Also, the target may not be stationary. In this case, the network can be trained with steady parameters and use the learned parameters only occasionally.

This leads to the *Deep Q-Network* (DQN) algorithm[e]:

---

[e]Never published in a peer-reviewed journal, but available as a pre-print: V. Mnih, K. Kavukcuoglu, D. Silver, A. Graves, I. Antonoglou, D. Wierstra, and M. Riedmiller, "Playing Atari with Deep Reinforcement Learning", arXiv:1312.5602 [cs.LG].

Algorithm 22: Deep Q-network

Input: state
for *each episode* **do**
$\quad$ ▷ Choose state s;
$\quad$ **repeat**
$\qquad$ ▷ Choose an action a derived from the ε-greedy algorithm;
$\qquad$ ▷ Take action a, obtain reward r and next state s';
$\qquad$ ▷ Add tuple (s, a, r, s') to the experience replay buffer;
$\qquad$ **Every** *so many steps* **do**
$\qquad\quad$ ▷ Sample mini-batch randomly from the experience replay
$\qquad\qquad$ buffer;
$\qquad\quad$ **for** *each tuple (s, a, r, s') in the mini-batch* **do**
$\qquad\qquad$ ▷ Compute expected value $y = r + \gamma \max\limits_{a' \in \mathcal{A}} Q(s', a')$;
$\qquad\qquad$ ▷ Train network with loss $\mathscr{L}(\theta) = \left\langle (y - Q_\theta(s, a))^2 \right\rangle$;
$\qquad\quad$ **end**
$\qquad\quad$ ▷ Update: $s \leftarrow s'$;
$\qquad$ **end**
$\quad$ **until** *a terminal state is reached*;
end

To create a replay buffer in Julia we use a structure that stores the buffer itself, its capacity, and the current position:

```julia
mutable struct ReplayBuffer
    buffer :: Vector{Tuple}
    max_size :: Int
    curr_size :: Int
end

function ReplayBuffer(buffer_size :: Int)
    return ReplayBuffer(Vector{Tuple}(), buffer_size, 0)
end
```

To store a tuple (s, a, r, s') in the buffer, we either include a new position or eliminate the first one circularly:

```julia
function store!(buffer :: ReplayBuffer, item)
    if (buffer.curr_size ≥ buffer.max_size)
        popfirst!(buffer.buffer)
    else
        buffer.curr_size += 1
    end
    push!(buffer.buffer, item)
end
```

A batch can be extracted easily with:

```julia
function sample(buffer :: ReplayBuffer, batch_size :: Int)
    indices = rand(1:length(buffer.buffer), min(batch_size,buffer.curr_size))
    return [buffer.buffer[i] for i in indices]
end
```

11 Gradient Methods

In the previous chapter, we began discussing reinforcement learning and introduced two important value methods: Q-learning and SARSA. In this chapter, we will continue our exploration of reinforcement learning redirecting our attention to directly optimizing the policy. We will study three key gradient-based methods: the REINFORCE algorithm, the actor-critic network, and trust-region methods. The first is one of the most basic algorithms that avoid the need for explicitly estimating the value function. While the REINFORCE algorithm is model-free and can handle stochastic policies, it suffers from high variance and does not explicitly address the exploration-exploitation trade-off. These factors can lead to slow convergence and stability. The actor-critic network, on the other hand, reduces the variance by incorporating a network that estimates the advantage of taking a specific action. In this architecture, the actor network explores the action space, while the critic evalutes the actions. This interplay allows for a balance between exploration and exploitation that can lead to faster convergences. The drawback of this configuration, however, is that it introduces more complexity that typically leads to more hyperparameter tuning. Completing this chapter, we will discuss trust-region methods. These strategies constrain the magnitude of policy updates within a trust region. This prevents large changes and can lead to better stability and convergence.

11.1 REINFORCE

Gradient policy methods are based on a *policy score function*, defined as the expected value of the cumulative reward over different trajectories [174]:

$$J = \langle R(\tau) \rangle_\tau = \sum_\tau \rho(\tau) R(\tau). \tag{11.1}$$

where τ is a trajectory $(s_0, a_0; \dots; s_{T-1}, a_{T-1}; s_T)$ generated by a policy π in an episode.

We would like to find the set of parameters that maximize this score function. For this, we take the gradient of this objective function for any parameter θ:

$$\nabla_\theta J = \sum_\tau \nabla_\theta \rho(\tau) R(\tau) = \sum_\tau \rho(\tau) \nabla_\theta \log\left(\rho(\tau)\right) R(\tau)$$
$$= \langle \nabla_\theta \log\left(\rho(\tau)\right) R(\tau) \rangle_\tau. \tag{11.2}$$

A Markov decision process is characterized by a sequence of tuples (s_t, a_t) [166]. The transitions between these tuples are given by a transition function:

$$P(s_{t+1}, a_{t+1} | \mathscr{F}_t) = P(s_{t+1}, a_{t+1} | s_t, a_t). \tag{11.3}$$

Given a policy function that determines which action to take given a specific state $\pi(a_t | s_t)$, we can use the definition of conditional probability to construct a Markov chain:

$$
\begin{aligned}
\rho(\tau) &= P(s_0, a_0; \ldots; s_T) \\
&= P(s_T | s_0, a_0; \ldots; s_{T-1}, a_{T-1}) P(s_0, a_0; \ldots; s_{T-1}, a_{T-1}) \\
&= P(s_T | s_{T-1}, a_{T-1}) P(a_{T-1} | s_0, a_0; \ldots; s_{T-1}) P(s_0, a_0; s_1, a_1; \ldots; s_{T-1}) \\
&= P(s_T | s_{T-1}, a_{T-1}) P(a_{T-1} | s_{T-1}) P(s_0, a_0; \ldots; s_{T-2}, a_{T-2}) \qquad (11.4) \\
&\cdots \\
&= P(s_0, a_0) \prod_{t=1} \pi(a_t | s_t) P(s_t | s_{t-1}, a_{t-1}).
\end{aligned}
$$

Taking its log:

$$
\log(\rho(\tau)) = \log(\rho(s_0, a_0)) + \sum_{t=1}^{T} \log(\pi(a_t | s_t)) + \sum_{t=1}^{T} \log(P(s_t | s_{t-1}, a_{t-1})). \quad (11.5)
$$

Now, taking its gradient:

$$
\nabla_\theta \log(\rho(\tau)) = \sum_{t=1}^{T} \nabla_\theta \log(\pi(a_t | s_t)). \quad (11.6)
$$

With this result, we can rewrite Eq. 11.2 as:

$$
\nabla_\theta J = \left\langle \sum_{t=1}^{T} \nabla_\theta \log(\pi(a_t | s_t)) R(\tau) \right\rangle_\tau. \quad (11.7)
$$

Observe that this is the maximum likelihood weighted by the cumulative reward of the episode.

Instead of explicitly calculating the expected value in the expression, it can be estimated by sampling trajectories. This leads to the *REINFORCE* algorithm summarized as [169, 174]:

Algorithm 23: REINFORCE

---

input: An analytic policy function $\pi_\theta(a_t | s_t)$
repeat
  ▷ Sample trajectory τ using the policy $\pi_\theta(a_t | s_t)$;
  ▷ Compute $\nabla_\theta J(\theta) \approx \sum_{t=1}^{T} \nabla_\theta \log(\pi_\theta(a_t | s_t)) R(\tau)$;
  ▷ Improve the policy with $\theta \leftarrow \theta + \alpha \nabla_\theta J(\theta)$;
until *Convergence*;

Whereas in Q-learning we needed a table, in REINFORCE we need to know an analytic policy function, which can be equally problematic. One solution, as found in the deep-Q algorithm, is to use a neural network. Particularly, in the REINFORCE algorithm, we can use a neural network with a softmax output since we need a probability mass function.

The reward for the trajectory $R(\tau_l)$ will only be known after the episode is complete. After this happens, the gradients have to be rescaled according to this reward.

11.1.1   A SIMPLE GAME

Let's consider a simple game to illustrate the algorithm. Here an agent is free to move on a 3×3 grid until she finds a treasure located at coordinates $(1,1)$ as shown in Fig. 11.1. The state of the agent is $\mathbf{s} = (s_1, s_2)$, and she can take any of nine actions: move up, down, left, right, up-left, up-right, down-left, and down-right.

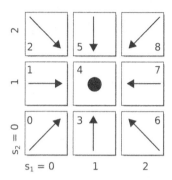

Figure 11.1   A simple game with expected actions illustrated at each site

For the policy, let's consider a model:

$$\pi_\theta(a|\mathbf{s}) = \exp\left\{-(a - \mu(\mathbf{s}))^2\right\}, \tag{11.8}$$

where

$$\mu(\mathbf{s}) = \omega_1 s_1 + \omega_2 s_2 + \omega_3. \tag{11.9}$$

According to the expected actions shown in Fig. 11.1, we see that when the state is $(0,0)$, the expected action is 0. Consequently, we find $\omega_3 = 0$. Likewise, when the state is $(0, s_2)$, the expected action is s_2. Therefore, $\omega_2 = 1$. Finally, when the state is $(2, s_2)$, the expected action is $y + 6$, which gives $\omega_1 = 3$. Thus, our model becomes:

$$\pi_\theta(a|\mathbf{s}) = \exp\left\{-(a - 3s_1 - s_2)^2\right\}. \tag{11.10}$$

To use the REINFORCE algorithm, we need:

$$\log(\pi_\theta(a_t|s_t)) = -(a - \mu(\mathbf{s}))^2$$

$$\frac{\partial}{\partial \omega_1} \log(\pi_\theta(a_t|s_t)) = 2(a - \omega_1 s_1 - \omega_2 s_2 - \omega_3) s_1$$

$$\frac{\partial}{\partial \omega_2} \log(\pi_\theta(a_t|s_t)) = 2(a - \omega_1 s_1 - \omega_2 s_2 - \omega_3) s_2 \qquad (11.11)$$

$$\frac{\partial}{\partial \omega_3} \log(\pi_\theta(a_t|s_t)) = 2(a - \omega_1 s_1 - \omega_2 s_2 - \omega_3).$$

With these results, we run the gradient ascent after each episode. After about 2×10^4 interactions, we obtain $\omega_1 \approx 2.955$, $\omega_2 \approx 1.033$, and $\omega_3 \approx 0.058$, which corresponds to an average story length of a single step.

This, however, is a very simple example. For more elaborated games, one typically uses a neural network with a softmax output as the policy. In this configuration, one of the outputs is chosen according to its probability, and the remaining ones are discarded. This can be obtained using a simple roulette selection algorithm [175]. This is a popular selection algorithm used, for example, in genetic algorithms (see Sec. 12.1). A vector with fitness values is the input of the function. One value between zero and the total fitness of the population is randomly chosen, and then, beginning from the first, add each individual's fitness until the desired value is obtained:

```
neural_output = FeedForward!(embedding(state),NeuralNetwork)
action_probabilities = softmax(neural_output)
action = roulette_selection(action_probabilities)

R += RewardMatrix[state,action]
```

In the snippet, "embedding" is a function that converts a state $(1,2,\ldots,N)$ to an input vector to the neural network ($\begin{bmatrix} 1 & 0 & 0 \end{bmatrix}$, for example).

To use this approach in the **REINFORCE** algorithm, we need the log of the softmax function:

$$\log(z_i(t)) = y_i(t) - \log\left(\sum_j e^{y_j(t)}\right). \tag{11.12}$$

The parameter θ shown in the **REINFORCE** algorithm is a parameter of the neural network (weights, biases, etc.). The derivative of the log of the softmax activation for such parameter is:

$$\frac{dL}{dy_k} = \frac{d\log(z_i)}{dy_k} = \delta_{ik} - \frac{e^{y_k(t)}}{\sum_j e^{y_j(t)}} = \delta_{ik} - z_k(t)$$
$$\frac{dL}{d\mathbf{y}(t)} = (\mathbf{1}_{ik} - \mathbf{z}(t))^\top \rightarrow \frac{dJ}{d\theta} = \sum_t \frac{dL}{d\mathbf{y}(t)}\frac{d\mathbf{y}(t)}{d\theta}R_t. \tag{11.13}$$

Note that although many outputs are discarded, they are used to calculate the gradient. In Julia:

```
    ⋮
    z = zeros(1,N)
    z[1,action] = 1
    ∇ -= z - reshape(action_probabilities,1,N)

    s = StateTransitionMatrix[state,action]
end

∇ *= R
BackPropagate(∇,NeuralNetwork)
SetGradients!(η,NeuralNetwork)
```

11.2 ACTOR-CRITIC

In the REINFORCE algorithm, it is necessary to compute the cumulative reward for the whole episode before upgrading the gradients of the network. Therefore, even trajectories taken at a time $t \neq 0$ need the rewards calculated in the past, which is counterintuitive. This problem can be circumvented with some algebra:

$$
\begin{aligned}
\nabla_\theta J(\theta) &= \left\langle \sum_{t=0}^{T-1} \nabla_\theta \log\left(\pi_\theta(a_t|s_t)\right) R(\tau) \right\rangle_\tau \\
&= \left\langle \sum_{t=0}^{T-1} \nabla_\theta \log\left(\pi_\theta(a_t|s_t)\right) \sum_{t'=0}^{T-1} \gamma^{t'} r(s_{t'}, a_{t'}, s_{t''}) \right\rangle_\tau \\
&= \sum_{t=0}^{T-1} \left[\left\langle \nabla_\theta \log\left(\pi(a_t|s_t)\right) \sum_{t'=0}^{t-1} \gamma^{t'} r(s_{t'}, a_{t'}, s_{t''}) \right\rangle_\tau + \right. \\
&\qquad \left. + \left\langle \nabla_\theta \log\left(\pi_\theta(a_t|s_t)\right) \sum_{t'=t}^{T-1} \gamma^{t'} r(s_{t'}, a_{t'}, s_{t''}) \right\rangle_\tau \right].
\end{aligned}
$$

The two terms inside the first brackets are independent since one depends on t and the other depends on the terms up to $t-1$. Therefore, we can write:

$$
\begin{aligned}
\nabla_\theta J(\theta) &= \sum_{t=0}^{T-1} \left[\langle \nabla_\theta \log\left(\pi(a_t|s_t)\right) \rangle_\tau \left\langle \sum_{t'=0}^{t-1} \gamma^{t'} r(s_{t'}, a_{t'}, s_{t''}) \right\rangle_\tau + \right. \\
&\qquad \left. + \left\langle \nabla_\theta \log\left(\pi_\theta(a_t|s_t)\right) \sum_{t'=t}^{T-1} \gamma^{t'} r(s_{t'}, a_{t'}, s_{t''}) \right\rangle_\tau \right].
\end{aligned}
\tag{11.14}
$$

The first expected value is zero, since for any distribution $P(\omega)$, we get:

$$
\begin{aligned}
&\int_\Omega P(\omega)d\omega = 1 \rightarrow \nabla_\theta \int_\Omega P(\omega)d\omega = 0 \\
&\int_\Omega P(\omega)\nabla_\theta \log\left(P(\omega)\right)d\omega = 0 \\
&\langle \nabla_\theta \log\left(P(\omega)\right)\rangle_\tau = 0.
\end{aligned}
\tag{11.15}
$$

Consequently, Eq. 11.14 can be written as:

$$
\nabla_\theta J(\theta) = \left\langle \sum_{t=0}^{T-1} \nabla_\theta \log\left(\pi_\theta(a_t|s_t)\right) \sum_{t'=t}^{T-1} \gamma^{t'-t} r(s_{t'}, a_{t'}, s_{t''}) \right\rangle_\tau,
\tag{11.16}
$$

where the second summation is often called *reward-to-go*. Also, convergence is faster when we use $t'-t$ as the exponent of γ. Now, only the future rewards are required to calculate the gradient of the objective function.

Another problem that appears in the REINFORCE algorithm is the high variance caused by sampling possibly many different trajectories. One way to circumvent this problem is to subtract a *baseline* from the cumulative reward. This has the effect of producing smaller gradients, which helps the algorithm to converge. We are allowed to do this as a direct consequence of Eq. 11.15 as long as the baseline does not depend

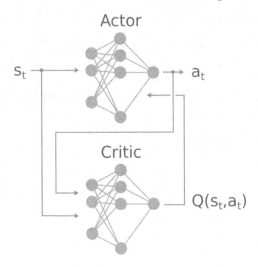

Figure 11.2   Actor-critic architecture

on the parameter θ. Since we are aiming at reducing variance, we can estimate the best baseline:

$$\sigma^2 = \left\langle [g(\tau)(R(\tau) - b)]^2 \right\rangle - \langle g(\tau)(R(\tau) - b) \rangle^2,$$

where $g(\tau) = \nabla_\theta \log(P_\theta(\tau))$.

Minimizing the variance, we obtain:

$$b = \frac{\langle g^2(\tau)R(\tau) \rangle_\tau}{\langle g^2(\tau) \rangle_\tau}. \tag{11.17}$$

A more common approach is to use the state-value function $V(s)$, which is the expected return for a policy. Thus, if the agent obtains the expected reward, a zero gradient is produced to correct the policy. From that arises the *actor-critic* approach, which utilizes two neural networks. One neural network is used to estimate the policy, while the other is used to estimate the state value function, as illustrated in Fig. 11.2 [169]. The loss function used in this new neural network is:

$$L(s) = V(s_t) - \sum_{t'=t}^{T-1} \gamma^{t'} r(s_{t'}, a_{t'}, s_{t''}). \tag{11.18}$$

Instead of using the reward-to-go in its current form, it is also possible to use the state-action function as we shall see. First, let's write $\hat{R} = \sum_{t'} \gamma^{t'} r(s_{t'}, a_{t'}, s_{t''})$ and split the trajectory into $\tau = \tau_{\rightarrow t}; \tau_{t \rightarrow T}$, where the first component corresponds to the trajectory up to instant t, and the second corresponds to the remaining of the trajectory up to the terminal state. The law of total expectation can now be used to write the gradient of the objective function as:

$$\nabla_\theta J(\theta) = \sum_{t=0}^{T-1} \left\langle \left\langle \nabla_\theta \log(\pi_\theta(a_t|s_t)) \hat{R} | \tau_{\rightarrow t} \right\rangle_\tau \right\rangle_\tau.$$

Because of the Markov property, we get:

$$
\begin{aligned}
\nabla_\theta J(\theta) &= \sum_{t=0}^{T-1} \left\langle \nabla_\theta \log\left(\pi_\theta(a_t|s_t)\right) \left\langle \hat{R}|s_t,a_t \right\rangle_\tau \right\rangle_\tau \\
&= \sum_{t=0}^{T-1} \left\langle \nabla_\theta \log\left(\pi_\theta(a_t|s_t)\right) Q(s_t,a_t) \right\rangle_\tau .
\end{aligned}
\tag{11.19}
$$

We can also subtract a baseline from the state-action function. A good choice again is to use the state-value function, which gives an average of the state.

Another possibility is to use the quantity $Q(s,a) - V(s)$, which is called the *advantage function* [176], and quantifies how better than average an action is. Hence, we end up with the following gradient for the objective function:

$$
\nabla_\theta J(\theta) = \left\langle \sum_{t=0}^{T} \nabla_\theta \log\left(\pi_\theta(a_t|s_t)\right) A(s_t,a_t) \right\rangle_\tau .
\tag{11.20}
$$

11.2.1   ADVANTAGE ACTOR-CRITIC

The advantage actor-critic (A^2C) algorithm [177] begins by taking an action with the actor-network. In Julia, we use a feedforward network with softmax output:

```
while(state ≠ final_state && steps < maxSteps)
    actor_output = actor(embedding(state))
    action = roulette_selection(actor_output)
```

In this snippet, the function embedding transforms an integer into a corresponding vector. For example, it transforms $state = 1$ into $\begin{bmatrix} 1 & 0 & 0 & 0 \end{bmatrix}^\top$. Function roulette_selection selects one of many outputs based on their probabilities (or fitnesses, as used in genetic algorithms):

```
function roulette_selection(fitness)
    total_fitness = sum(fitness)
    r = rand()*total_fitness

    i = 0
    while(r ≥ 0)
        i += 1
        r -= fitness[i]
    end

    return i
end
```

After an action is chosen, we obtain the next state and the reward:

```
r = RewardMatrix[state,action
next_state = StateTransitionMatrix[state,action]
```

Next, we need to use an adequate baseline. The REINFORCE loss depends on summing the gradients of the whole episode, and then weighting it by the cumulative reward of the whole episode. We can do some algebra to avoid this problem:

$$\nabla_\theta J(\theta) = \left\langle \sum_{t=0}^{T-1} \nabla_\theta \log(\pi_\theta(a_t|s_t)) R_t \right\rangle_\tau$$

$$= \sum_{t=0}^{T-1} \langle \nabla_\theta \log(\pi_\theta(a_t|s_t)) \rangle_\tau \langle R_t \rangle_\tau. \tag{11.21}$$

The value $\langle R_t \rangle_\tau$ is just the Q-value defined in Eq. 10.7, but instead of explicitly calculating it, we will estimate it with a neural network. Thus, we have:

$$\nabla_\theta J(\theta) = \left\langle \sum_{t=0}^{T-1} \nabla_\theta \log(\pi_\theta(a_t|s_t)) Q_w(s_t,a_t) \right\rangle_\tau, \tag{11.22}$$

where $Q_w(s_t,a_t)$ is the estimated Q-value from a neural network.

The parameters of the actor network are updated directly from Eq. 11.22, whereas time-difference is used to update the critic network:

$$\delta_t = r_t + \gamma Q_w(s',a') - Q_w(s,a)$$

$$w \leftarrow w + \alpha_w \delta_t \Delta_w Q_w(s,a). \tag{11.23}$$

Since we are Monte-Carlo sampling trajectories in the REINFORCE algorithm, trajectories can considerably deviate from one another during training, and this causes high variability in log probabilities. This high variability, in turn, produces noisy gradients that can make learning difficult and sub-optimal. Moreover, policies that produce a zero cumulative reward do not help to distinguish good from bad policies.

As already discussed, one way to reduce these problems is by subtracting a baseline from the cumulative reward:

$$\nabla_\theta J(\theta) = \left\langle \sum_{t=0}^{T-1} \nabla_\theta \log(\pi_\theta(a_t|s_t)) (R_t - b(s_t)) \right\rangle_\tau. \tag{11.24}$$

This has the effect of producing smaller gradients and, consequently, more stable updates. Now, instead of using two networks to estimate both Q and V_ϕ, we can use the Bellman optimality equation (Eq. 10.9) to write:

$$Q(s_t,a_t) = \langle r_{t+1} + \gamma V_\phi(s_{t+1}) \rangle. \tag{11.25}$$

Therefore, the advantage value can be written as:

$$A(s_t, a_t) = r_{t+1} + \gamma V_\phi(s_{t+1}) - V_\phi(s_t), \qquad (11.26)$$

where V_ϕ is estimated by the critic. Observe that the temporal difference error is an unbiased estimate of the advantage function (see Sec. 10.2). In Julia, we calculate the advantage as:

```
lastV = V
V = critic(embedding(next_state))[1,1]
A = r + γV - lastV
```

The gradient with an advantage function becomes:

$$\nabla_\theta J(\theta) = \left\langle \sum_{t=0}^{T-1} \nabla_\theta \log\left(\pi_\theta(a_t|s_t)\right) A(s_t, a_t) \right\rangle_\tau . \qquad (11.27)$$

To implement this in Julia, we can use the Flux library again:

```
∇a = Flux.gradient(params(actor)) do
    -sum(log.(actor(embedding(s))).*Flux.onehot(action, 1:num_of_states))*A
end

Flux.Optimise.update!(actor_optimizer,params(actor),∇a)
```

The actor network produces an output with probabilities for each action, but we are only interested in the one that was taken. This selection is made by multiplying the chosen action by one and the remaining ones by zero. The vector that implements this is given by the function "one-hot." The result is passed as an argument to the function "gradient," which also uses the actual parameters of the network. These parameters are then updated based on this gradient using some optimizer.

The loss used in the critic network is related to the difference between the estimated $V^\pi(s_t)$ and the target $r_t + \gamma V^\pi(s_{t+1})$ state-value functions:

```
∇c = Flux.gradient(params(critic)) do
    Flux.mse(critic(embedding(state)),r+γ*V)
end

Flux.Optimise.update!(critic_optimizer,params(critic),∇c)
```

With this, we arrive at the online A^2C learning algorithm [177]:

Algorithm 24: A^2C

---

input: Actor policy model π_θ and critic value model V_ϕ
repeat
  ▷ Run policy $\pi_\theta(\mathbf{a}|\mathbf{s})$ using the actor network to get $(\mathbf{a}, r, \mathbf{s}')$;
  ▷ Compute $V_\phi(\mathbf{s}')$ using the critic network and find the advantage
  $A = r + \gamma V_\phi(\mathbf{s}') - V_\phi(\mathbf{s})$;
  ▷ Update critic $V_\phi(\mathbf{s})$ with the target $r + \gamma V_\phi(\mathbf{s}')$;
  ▷ Update actor with the gradient $\nabla_\theta \log(\pi_\theta(\mathbf{a}|\mathbf{s})) A(\mathbf{s}, \mathbf{a})$;
until *convergence*;

Note how the policy gradient is always written as $\langle \nabla_\theta \log(\pi_\theta(a_t|s_t)) \varphi_t \rangle$. If φ_t weights more rewards, then the gradient is on average more correct at the cost of high variance. On the other hand, if φ_t weights more estimation, then it is more stable at the cost of high bias. Therefore, choosing an appropriate φ function may lead to an optimal bias × variance relationship.

For a n-step TD, we have:

$$A_t^n = \sum_{k=0}^{n-1} \gamma^k r_{t+k} + \gamma^n V(s_{t+n}) - V(s_t)$$

$$A_t^1 = r_t + \gamma V(s_{t+1}) - V(s_t) = \delta_t$$

$$A_t^2 = r_t + \gamma r_{t+1} + \gamma^2 V(s_{t+1}) - V(s_t) = \delta_t + \gamma \delta_{t+1}$$

A generalized advantage estimator (GAE)[a] can be defined as the exponentially-weighted average:

$$\begin{aligned}
A_t^{\text{GAE}} &= (1 - \lambda)\left(A_t^1 + \lambda A_t^2 + \lambda^2 A_t^3 + \ldots\right) \\
&= (1 - \lambda)\left(\delta_t + \lambda(s_t + \gamma \delta_{t+1}) + \ldots\right) \\
&= (1 - \lambda)\left(\delta_t(1 + \lambda + \lambda^2 + \ldots) + \delta_{t+1}\gamma\lambda(1 + \lambda + \lambda^2 + \ldots) + \ldots\right) \quad (11.28) \\
&= \sum_{k=0}^{\infty} \delta_{t+k}(\gamma\lambda)^k,
\end{aligned}$$

where λ allows us to find an appropriate bias × variance relationship.

Thus, the gradient can be written as:

$$\nabla_\theta J(\theta) = \left\langle \nabla_\theta \log(\pi_\theta(s_t, a_t)) \sum_{k=0}^{\infty} \delta_{t+k}(\gamma\lambda)^k \right\rangle. \quad (11.29)$$

---

[a]Never published in a peer-reviewed journal, but available as a pre-print: J. Schulman, P. Moritz, S. Levine, M. I. Jordan, and P. Abbeel, "High-Dimensional Continuous Control Using Generalized Advantage Estimation" arXiv:1506.02438 [cs.LG].

11.3 TRUST REGION METHODS

The line search [178] is an algorithm widely used in optimization problems. In a line search, a direction that brings the objective function closer to zero is found, and then a step of a specific size is taken in this direction. In contrast, in a trust region algorithm [179], we first establish a step size and use it as the radius of a region[b]. To proceed, a point that minimizes our objective function within this region is found. This minimization provides the new direction for the algorithm, guiding its subsequent iterations. A visual representation of this process is illustrated in Fig. 11.3.

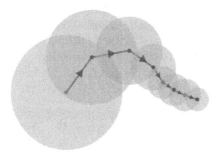

Figure 11.3 A typical output of the trust region algorithm

In the policy gradient algorithms, we update the policy with first-order derivatives. If the objective surface has a high curvature, though, we may end up producing bad moves causing instability or stopping in a suboptimal solution. It is possible to mitigate this problem by keeping the new and old policies close in parameter space. This can be accomplished with trust regions, and to implement this in reinforcement learning, we can maximize the policy score function subject to a maximum step size in parameter space. The latter can be done by ensuring that the old and new policies are close to each other. In other words, we must keep the Kulback-Leibler divergence between policies small:

$$\Delta^* = \arg \max_{\Delta} J(\theta_0 + \Delta) - \lambda \left(D_{KL}(\pi_{\theta_0} \| \pi_\theta) - \varepsilon \right), \qquad (11.30)$$

where ε is the trust radius or the maximum tolerable divergence between policies, θ_0 is an original parameter, and $\Delta = \theta - \theta_0$ is a step toward the new policy.

Using a first-order Taylor expansion for the score function and the Fisher information matrix as an approximation for $D_{KL}(\pi_{\theta_0} \| \pi_\theta)$ (see Sec. 2.2.1), we can write the following Lagrange function:

$$\mathcal{L} = -\nabla_\theta J(\theta)|_{\theta=\theta_0} \cdot \Delta + \lambda \left(\frac{1}{2} \Delta^\top \mathbf{H}_D(\theta_0) \Delta - \varepsilon \right), \qquad (11.31)$$

---

[b]For example, the Levenberg-Marquardt algorithm (see Sec. 6.5) takes a similar approach by restricting the update to a small region: $\lambda \, \text{diag}(\mathbf{J}^\top \mathbf{J})$.

where λ is a Lagrange multiplier. Minimizing this Lagrangean as a function of Δ, we get:

$$\frac{\partial \mathscr{L}}{\partial \Delta} = -\nabla_\theta J(\theta)|_{\theta=\theta_0} + \lambda \mathbf{H}_D(\theta_0)\Delta = 0$$

$$\Delta = \frac{1}{\lambda}\mathbf{H}_D(\theta_0)^{-1}\nabla_\theta J(\theta)|_{\theta=\theta_0} \tag{11.32}$$

$$= \tilde{\nabla}_\theta J(\theta)|_{\theta=\theta_0}.$$

In this expression, $\tilde{\nabla}_\theta$ represents the *natural gradient* [180, 181], which derives its name from the Fisher information matrix. This matrix characterizes the curvature of the loss function with respect to the model parameters. It offers a more "natural" method for transversing the probability distribution space compared to the standard gradient. Consequently, the natural gradient offers a more effective exploration of the parameter space in terms of probability distributions.

The Lagrange multiplier can be computed using the Lagrange duality [87]. Using a simplified notation where $\mathbf{g}(\theta_0) = \nabla_\theta J(\theta)|_{\theta=\theta_0}$, we get:

$$\mathscr{L} = -\frac{1}{\lambda}\mathbf{g}(\theta_0)\mathbf{H}^{-1}(\theta_0)\mathbf{g}(\theta_0) + \frac{1}{2\lambda}\mathbf{g}^\top(\theta_0)\mathbf{H}^{-1}(\theta_0)\mathbf{g}(\theta_0) - \lambda\varepsilon. \tag{11.33}$$

Maximizing it for λ, we get:

$$\lambda = \sqrt{\frac{\mathbf{g}^\top(\theta_0)\mathbf{H}^{-1}(\theta_0)\mathbf{g}(\theta_0)}{2\varepsilon}}. \tag{11.34}$$

Consequently, the optimal step size in parameter space is:

$$\Delta = \sqrt{\frac{2\varepsilon}{\mathbf{g}^\top(\theta_0)\mathbf{H}^{-1}(\theta_0)\mathbf{g}(\theta_0)}}\mathbf{H}^{-1}(\theta_0)\mathbf{g}(\theta_0). \tag{11.35}$$

Computing the inverse of the Fisher information matrix, however, is expensive and techniques such as conjugate gradient are usually used (see Appendix H). This leads to the following trust region policy optimization (TRPO) algorithm [182]:

Algorithm 25: TRPO

---
    input: An analytic policy function $\pi_\theta(a_t|s_t)$
    repeat
        $\triangleright$ Collect a set of trajectories by running the policy π_θ;
        $\triangleright$ Compute the rewards-to-go and the advantage estimates;
        $\triangleright$ Estimate policy gradient: $\nabla_\theta J(\theta) \approx \sum_{t=0}^{T-1}\nabla_\theta \log(\pi_\theta(a_t|s_t))A(t)$;
        $\triangleright$ Use conjugate gradient to compute the inverse of the Fisher information matrix;
        $\triangleright$ Update the parameters of the policy using Eq. 11.35;
        $\triangleright$ Update the value function;
    until *Convergence*;

The constraint on the KL divergence may be violated considering all the approxima-tions. To avoid this problem, the parameters are typically updated using a line search that effectively shrinks the trust region:

Algorithm 26: TRPO Line Search

---

input: Δ computed using Eq. 11.35, and $\alpha \in (0,1)$
for $j = 0,\ldots,L$ **do**
 $\triangleright$ Compute $\theta = \theta_0 + \alpha^j \Delta$;
 if $J(\pi_\theta) \geq J(\pi_{\theta_0})$ *and* $D_{KL}(\theta_0 \| \theta) \leq \varepsilon$ **then**
  $\triangleright$ Accept the update and make $\theta = \theta_0 + \alpha^j \Delta$;
  $\triangleright$ break;
 end
end

TRPO involves complex mathematics, making it challenging to implement. How-ever, Proximal Policy Optimization (PPO)[c] solves this problem by employing clip-ping, as depicted in Fig. 11.4, to facilitate small policy updates. Unlike TRPO, PPO does not rely on restricting the KL divergence or computing the Fisher information matrix, which is beneficial. If the advantage is positive, the action is better than ex-pected and should have a high probability of being used again. However, to avoid a high-importance sampling weight, the objective is clipped, maintaining the distance between policies small. The same is valid for a negative advantage but in the opposite direction. Thus, the objective to be minimized becomes:

$$\mathscr{L}^{\text{clip}}(\theta) = \left\langle \min\left(r_\theta(t)\hat{A}_t, \text{clip}\left(r_\theta(t), \frac{1-\varepsilon}{1+\varepsilon} \right) \hat{A}_t \right) \right\rangle, \qquad (11.36)$$

where:

$$r_\theta(t) = \frac{\pi_\theta(a_t|s_t)}{\pi_{\theta_0}(a_t|s_t)} \qquad (11.37)$$

is a probability ratio between the action under the new and old policies, and $\varepsilon \in [0.1, 0.3]$ is a hyperparameter. This can be easily implemented in Flux/Julia as:

```
∇a .+= Flux.gradient(params(actor)) do
    πθ = actor(embedding(s))·Flux.onehot(action,1:5)
    r = exp(log(πθ) - log(πθ₀))
    -clip(A,ε,r)
end
```

---

[c]Never published in a peer-reviewed journal, but available as a pre-print: J. Schulman, F. Wolski, P. Dhariwal, A. Radford, O. Klimov, "Proximal Policy Optimization Algorithms" arXiv:1707.06347 [cs.LG].

Note that: i) instead of calculating the quotient A/B, we calculate $\exp(\log(A) - \log(B))$ because it is numerically more stable, and ii) we need the negative of the gradient because we are computing a gradient ascent, instead of descent.

We can improve the algorithm by turning it off-policy. To do this, we use importance sampling (see Appendix I), which consists of sampling from an old distribution. From the policy score function, we can write:

$$J(\theta) = \sum_\tau \pi_\theta(\tau)A(\tau) = \sum_\tau \pi_{\theta_0}(\tau)\frac{\pi_\theta(\tau)}{\pi_{\theta_0}(\tau)}A(\tau) = \left\langle \frac{\pi_\theta(\tau)}{\pi_{\theta_0}(\tau)}A(\tau) \right\rangle_{\tau_0}, \qquad (11.38)$$

where the probability ratio appears naturally in the equation. Thus, the policy score function can be computed following the old policy, storing values, and then using them to update the parameters of the actor and critic networks.

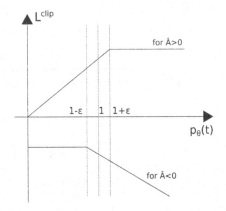

Figure 11.4 Clipping function used in the PPO algorithm

With this, we arrive at the PPO algorithm:

Algorithm 27: PPO

---

repeat
  for *each of N actors in parallel* **do**
      ▷ Collect T transitions using the actor policy $\pi_\theta(a_t|s_t)$;
      ▷ Estimate the T advantages using the critic:
      $A(s_t, a_t) \approx r(s_t, a_t) + \gamma V_\phi(s_{t+1}) - V_\phi(s_t)$;
  end
  for *each epoch* **do**
      ▷ Sample $M \leq NT$ transitions from the collected ones;
      ▷ Calculate the gradient of the clipped objective;
      ▷ Compute critic loss;
  end
  ▷ Update actor $\theta \leftarrow \theta + \alpha\nabla J(\theta)$;
  ▷ update critic $\phi \leftarrow \phi - \beta\nabla L(\phi)$;
until *Convergence*;

Section VI

Optimization

12 Population-Based Metaheuristic Methods

Evolutionary computing [183, 184] is a technique that mimics the natural selection process found in biological systems to find the best solutions for a specific problem. It achieves this by evolving a population of potential solutions over several generations, with the fittest individuals having a higher chance of producing the next generation. Compared to gradient-based optimization methods, evolutionary approaches bypass the need for gradient computations, resulting in faster and more efficient performances. Additionally, they offer high parallelizability, allowing for concurrent processing, and have the ability to generate a list of solutions rather than a single one. Furthermore, this list can be continually improved over time, making evolutionary computing a valuable tool for solving complex problems.

In general, evolutionary algorithms follow a set of steps: i) a random population is formed, ii) a fitness function is evaluated for each individual, iii) the fittest individuals are selected for reproduction, iv) mutation and crossover operations are performed to produce the new generation, and v) the individuals with the least fitness functions are replaced. This is illustrated in Fig. 12.1.

In this chapter, we will discuss some of these approaches such as genetic algorithms, particle swarm optimization, and ant colony optimization.

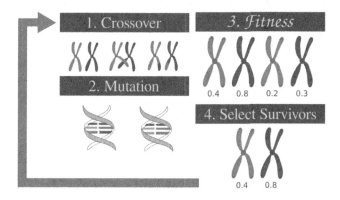

Figure 12.1   Fluxogram of a genetic algorithm

12.1 GENETIC ALGORITHMS AND EVOLUTION STRATEGY

In genetic algorithms (GA) [183], a population is a set of solutions, a chromosome is a particular solution among this pool, a gene is a particular position of the chromosome array, and an allele is the value of that gene. This is illustrated in Fig. 12.2.

Typically, genes carry binary values, whereas evolution strategies (ES) consider real-valued genes. For example, ES has been used in solving inverse problems in quantum mechanics. In this approach, a population is iteratively improved until it converges to a configuration that closely resembles the measured quantum state [185, 186].

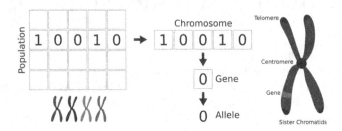

Figure 12.2 Structure of a chromosome: sister chromatids are united through a centromere. The end caps of a chromatid is a telomere and chromatids are composed of genes

Most genetic algorithms use the concept of Darwinian[a] evolution [187] where species evolve through natural selection over time. Individuals that are better adapted are more likely to survive and reproduce, passing their advantageous traits to their offspring. Some GA can incorporate other evolutionary approaches [188]. For example, Lamarckian[b] evolution states that individuals can pass on traits acquired during their lifetime to their offspring, and Baldwinian[c] evolution proposes that evolutionary changes can occur as a result of the interaction between the individual's genes and the environment. Next, we will discuss each step of these algorithms.

12.1.1   SELECTION

Care should be taken to keep diversity in the population or there can be a premature convergence. In this situation the entire population is taken by one solution. To circumvent this issue, one can randomly pick individuals for mating. This, however, is usually avoided because no selection criterium is imposed. Rather, strategies such as tournament, roulette wheel, and stochastic sampling are used.

12.1.1.1   k-Way Tournament

In a k-way selection [189, 190], k chromosomes are picked randomly. Among these k chromosomes, the one with the highest fitness function is chosen with probability p, the next with probability $p(1-p)$, the next with probability $p(1-p)^2$, and so on. If $p = 1$, then the selection is deterministic, and the element with the highest fitness value among the k competitors is always selected. Also, if $k = 1$, then any individual

---

[a]Charles Robert Darwin (1809–1882), English biologist, winner of many prizes, including the Copley Medal in 1879.
[b]Jean-Baptiste Lamarck (1744–1829), French biologist.
[c]James Mark Baldwin (1861–1934), American philosopher.

is randomly chosen from the entire population. The k-way procedure for $k = 3$ is shown in Fig. 12.3.

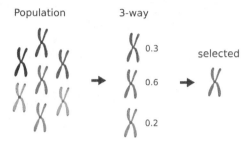

Figure 12.3 Three-way tournament procedure example

The deterministic k-way tournament can be implemented in Julia as:

```julia
function kway(Fitness,k,n)
    N = length(Fitness)

    selection = []
    for i in 1:n
        idx = rand(1:N,k)
        p = argmax(Fitness[idx])
        push!(selection,idx[p])
    end

    return selection
end
```

12.1.1.2 Roulette Wheel Selection

In a roulette wheel selection (or *fitness proportionate selection* – FPS) [175], probabilities are assigned to each chromosome as:

$$p_i = \frac{\mathscr{F}_i}{\sum_{n=1}^{N} \mathscr{F}_i}, \tag{12.1}$$

where N is the size of the population, and $\mathscr{F}$ is the fitness function.

The probabilities are equivalent to areas of a roulette. Thus, in an analogy, the roulette would spin and the selected area would correspond to the chosen chromosome. In real computations, a cumulative probability function is computed, and a uniformly distributed random number is chosen. The inverse CDF for this number gives the chromosome. This procedure is illustrated in Fig. 12.4.

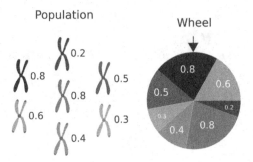

Figure 12.4   Population with fitness values and corresponding wheel

```
function roulette_selection(fitness)
    total_fitness = sum(fitness)
    r = rand()*total_fitness

    i = 0
    while(r ≥ 0)
        i += 1
        r -= fitness[i]
    end

    return i
end
```

Another possibility is to use the alias method. Here, we use a probability table U_i and an alias table K_i with entries $1 \leq i \leq n$. We then pick a random variable $x \in [0;1)$, create the table index $i = \lfloor n.x \rfloor + 1$, and set $y = n.x + 1 - i$. Consequently, i becomes uniformly distributed on $[1;n]$ and y becomes uniformly distributed on $[0;1)$. If $y < U_i$, then the algorithm returns i, otherwise, it returns K_i.

An even simpler method, *stochastic acceptance* [191], consists of, instead of searching, selecting a chromosome and accepting it with probability $\mathscr{F}_i/\mathscr{F}_M$, where $\mathscr{F}_M$ is the maximum fitness in the population. The process is repeated until a chromosome is chosen.

FPS has the main disadvantage that, if there is an individual with a relatively large fitness function, the remaining individuals may not be chosen. There is a variation of the FP selection where many individuals are chosen simultaneously using evenly spaced pointers. This technique, known as Stochastic Universal Selection (SUS) [192], can overcome the main problem of FPS.

12.1.2 CROSSOVER

During meiosis, paired homologs, which consist of both paternal and maternal chromosomes that are non-sister chromatids of the same type, undergo a series of events. They align through synapsis, experience breakage, exchange segments, and finally separate. This process, called crossover [183], may occur more than once with the same pair, producing diversity.

In genetic algorithms, this corresponds to several strategies. It is possible, for instance, to randomly select one or multiple breakpoints and recombine genes, or we can pick randomly alleles from each gene with a certain probability. The first two strategies are called one and multi-point crossover, while the last strategy is called uniform crossover. It is also possible to perform an arithmetic recombination, typically given by:

$$z_i = \alpha_i.x + (1 - \alpha_i).y. \tag{12.2}$$

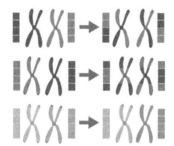

Figure 12.5 (Top) single point crossover, (middle) multi-point crossover, and (bottom) uniform crossover

12.1.3 MUTATION

As pointed out earlier, mutation is necessary to maintain the diversity of the population, which improves the convergence of the algorithm. Some mutation operators include [183]:

i) Single gene mutation, which flips an allele if it is digital or randomly assigns it if it has an integer representation.

ii) Swap mutation, which swaps the alleles of two randomly chosen genes.

iii) Scramble mutation, which scrambles the alleles of a genetic subset.

iv) Inversion mutation, which inverts the alleles of a genetic subset.

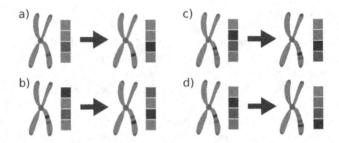

Figure 12.6 (a) single gene mutation, (b) swap mutation, (c) scrabble mutation, and (d) inversion mutation

In the case of the evolutionary strategy[d] [193], other mutation strategies are used such as adding a value sampled from a specific distribution or introducing correlations between the elements of the candidate solutions.

12.1.4 SURVIVOR SELECTION AND TERMINATION

One could consider randomly selecting individuals to be taken off the gene pool. This, however, suffers from convergence problems. Typically, elitist and age approaches are used. In the former, individuals with the lowest fitness function are discarded. The selection itself can be made with k-way, roulette wheel, and the other techniques previously discussed. For age-based selection, all individuals are allowed to stay a finite duration in the game and try to reproduce. After that, the individuals are discarded regardless of their fitness functions. The simulation terminates if there is no improvement in the population after a few interations, if a predetermined number of generations have passed, or if the objective function has reached some value.

12.2 PARTICLE SWARM OPTIMIZATION

In particle swarm optimization [194], a collection of agents explore the parameter space, communicating with one another to find the best configuration. In a simulation step, the position $\mathbf{x}_i$ of each particle "i" is updated according to:

$$\mathbf{x}_{i,t+1} = \mathbf{x}_{i,t} + \mathbf{v}_{i,t+1}\Delta_t, \tag{12.3}$$

where Δ_t is the time interval between indexes t and $t + 1$.

The velocity $\mathbf{v}_i$ of each individual is updated based on their own experience regarding their best position and the best position discovered by the group. The first component is related to a cognitive capability (c_1), whereas the second is related to

---

[d]Developed by Ingo Rechenberg (1934–2021) and Hans-Paul Schwefel (1940–Present) German computer scientists.

a social ability (c_2):

$$\mathbf{v}_{i,t+1} = w_t \mathbf{v}_{i,t} + c_1 \frac{r_1}{\Delta_t}\left(\mathbf{p}_{i,t} - \mathbf{x}_{i,t}\right) + c_2 \frac{r_2}{\Delta_t}\left(\mathbf{p}_{g,t} - \mathbf{x}_{i,t}\right). \tag{12.4}$$

Here, w_t is a component related to the inertia of the particle, usually between 0.4 and 1.0 to prevent divergence. Elements $r_{1,2}$ are random numbers uniformly distributed between 0 and 1 that help the agents explore the parameter space. Element $\mathbf{p}_i$ is the best individual solution, whereas $\mathbf{p}_g$ is the best solution found by the swarm up to a specific time step. Parameter c_1 controls how much importance the agent gives to his own experience, allowing him to explore the search space more aggressively. Parameter c_2, on the other hand, encourages the exploitation of regions that have been successfully explored by other particles. Both coefficients are empirically chosen in the range between 0.0 and 4.0. Many simulations begin with $c_1 = c_2 = 2.0$, which sets equal importance to the agent's own experience and the global experience.

A simulation proceeds according to the algorithm below:

Algorithm 28: Particle Swarm Optimization

---

repeat

    Compute fitness $\mathscr{F}_{i,t}$ of each individual at instant t;

    if ($\mathscr{F}_{i,t}$ *is better than* $\mathscr{F}_{i,best}$) **then**

        $\mathscr{F}_{i,best} = \mathscr{F}_{i,t}$

        $\mathbf{p}_{i,t} = \mathbf{x}_{i,t}$

    end

    if ($\mathscr{F}_{i,t}$ *is better than* $\mathscr{F}_{g,best}$) **then**

        $\mathscr{F}_{g,best} = \mathscr{F}_{i,t}$

        $\mathbf{p}_{g,t} = \mathbf{x}_{i,t}$

    end

    Update velocities using:

        $\mathbf{v}_{i,t+1} = w_t \mathbf{v}_{i,t} + c_1 \frac{r_1}{\Delta_t}\left(\mathbf{p}_{i,t} - \mathbf{x}_{i,t}\right) + c_2 \frac{r_2}{\Delta_t}\left(\mathbf{p}_{g,t} - \mathbf{x}_{i,t}\right)$;

    Update positions using:

        $\mathbf{x}_{i,t+1} = \mathbf{x}_{i,t} + \mathbf{v}_{i,t+1}\Delta_t$;

until *stop criterium has been reached*;

Optimization algorithms can be tested using non-convex[e] functions such as the Rosenbrock[f] and the Rastrigin[g] functions:

$$f_{Rosen}(x,y) = (a-x)^2 + b(y-x^2)^2$$
$$f_{Rast}(x,y) = 2A + x^2 + y^2 - A\left[\cos(2\pi x) + \cos(2\pi y)\right]. \tag{12.5}$$

---

[e]A function f is non-convex if a point in a straight line segment between any two points on the function can lie below the function.

[f]Howard Harry Rosenbrock (1920–2010) English engineer.

[g]Leonard Andreevich Rastrigin (1929–1998) Russian engineer.

The Rosenbrock function has a global minimum at (a, a^2), and typical values are $a = 1$, and $b = 100$. The Rastrigin function has a global minimum at $(0,0)$. In most cases, $A = 10$, and the search domain is defined as $x, y \in [-5.12, 5.12]$.

Figure 12.7 shows three different simulation steps of a swarm of 25 particles finding the global minimum of the Rastrigin function. It takes an average of 100 steps to achieve a precision of 10^{-5} using $c_1 = 1.5$, $c_2 = 0.8$, and $w = 0.8$.

Figure 12.7   Three PSO steps for 25 particles progressively finding the global minimum for the Rastrigin function

12.3   STIGMERGY ALGORITHMS

Stigmergy[h] is an indirect communication mechanism used by agents with limited awareness of each other, achieved through environmental manipulation. One of the most popular algorithms that try to mimic this biological behavior is the ant colony optimization[i] [195], which we will describe here. In this algorithm, ants communicate with each other by leaving a pheromone trail that reinforces the path traveled by other members of the colony. This creates a positive feedback loop that favors a particular type of solution. The structure of the algorithm is very simple. During each simulation step, solutions are generated, and additional procedures can be optionally employed to guide the search process. Finally, the pheromone levels are updated. This cycle continues until some stopping criterium is reached.

In the step of constructing solutions, the ants move stochastically through the nodes of a *construction graph*[j] $G = (V, E)$ without moving to positions previously traveled. This procedure continues until a target node is reached. The probability of moving from node i to a node j is given by:

$$p_{ij} = \frac{\tau_{ij}^{\alpha} \eta_{ij}^{\beta}}{\sum_{z \in \mathcal{N}(i)} \tau_{iz}^{\alpha} \eta_{iz}^{\beta}}, \tag{12.6}$$

where $\alpha, \beta \in \mathbb{R}_+$ are hyperparameters related to the importance of pheromone and heuristic information, respectively. Parameter τ is a matrix containing the pheromone

---

[h]Introduced by Pierre-Paul Grassé (1895–1985) French zoologist.
[i]Proposed by Marco Dorigo (1961–Present) Italian systems scientist.
[j]see Sec. 4.4 and Appendix B.

level in the edges, and η is a heuristic value associated with the edges, usually denoting the inverse of the distance between nodes. $\mathcal{N}(i)$ refers to the neighborhood of node i.

After all ants have constructed their solutions, the pheromone levels are updated according to:

$$\tau_{ij} = (1 - \rho)\tau_{ij} + \sum_k \Delta_{ij}(z), \qquad (12.7)$$

where $\rho \in (0, 1]$ is the pheromone evaporation rate, and $\Delta_{ij}(z)$ is the amount of pheromone deposited on edge e_{ij} by the kth ant. This is, typically, given by:

$$\Delta_{ij}(k) = \begin{cases} Q/L_k & \text{if edge } e_{ij} \text{ is in the path traveled by ant } k, \\ 0 & \text{otherwise.} \end{cases} \qquad (12.8)$$

In this last equation, Q is a fixed amount of pheromone, and L_k is the length of the path traveled by ant k.

To implement the ant colony algorithm in Julia, we create two mutable structures, one for an ant containing its position and memory, and another one for the colony containing the hyperparameters, the adjacency matrix, the pheromone level matrix, and the ants:

```
mutable struct antAgent
      node        :: Int
      memory   :: Vector
end

mutable struct antColony
      number_of_ants   :: Int
      target_node        :: Int
      τ                        :: Matrix
      A                        :: Matrix
      ants                    :: Vector
      α                        :: Float64
      β                        :: Float64
      ρ                        :: Float64
      Q                        :: Float64
end
```

To build solutions, first we need to find the neighboring nodes and avoid those already visited:

```
neighbors = find_neighbors(colony.A,ant.node)
neighbors = setdiff(neighbors,ant.memory)
```

We then find the probabilities of moving to each of these nodes:

```
for j in neighbors
    η = 1.0/colony.A[ant.node,j]
    p = (colony.τ[ant.node,j]^colony.α)*(η^colony.β)
    push!(fitness,p)
end
```

Finally, we move to a new node and store it in the memory of traveled nodes:

```
k = neighbors[roulette_selection(fitness)]
setfield!(ant,:node,k)

memory = ant.memory ∪ k
setfield!(ant,:memory,memory)
```

Figure 12.8 shows the result of a few simulation steps to find the shortest path between nodes 1 and 5 in a simple graph.

There are several variations of the ant colony algorithm. The one we have discussed is referred to as the *ant system* algorithm [196]. Another variation is the *ant colony system* [197], which uses a pseudorandom proportional rule. In this algorithm, the ant moves stochastically to a new node only if a random variable, uniformly distributed over the interval $(0,1]$, is greater than a threshold value q_0. If the random variable is less than or equal to q_0, the ant greedily moves to the node that maximizes the product τ_{ij}^α and η_{ij}^β. This strategy favors the exploitation of the solution space. Moreover, during the first step of the algorithm, the local pheromone is updated as:

$$\tau_{ij} = (1 - \rho)\tau_{ij} + \rho\tau_0, \tag{12.9}$$

where τ_0 is the initial value of the pheromone level.

Another variation is the *max-min ant system* [198]. In this algorithm, only the ant that produced the best solution adds a pheromone to τ:

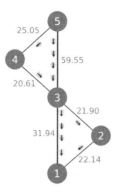

Figure 12.8 A simple graph of five nodes and the pheromone level per edge after applying a few steps of the ant colony optimization algorithm. The goal of the colony is to find the shortest route between nodes 1 and 5

$$\tau_{ij} = (1 - \rho)\tau_{ij} + \Delta_{ij}^{\text{best}}. \tag{12.10}$$

Furthermore, the pheromone level is clamped between minimum and maximum values.

13 Local Search Methods

In the previous chapter, we focused on population-based metaheuristics. These are robust techniques that can handle complex problems and find global optimal or near-optimal solutions. This happens at the cost of high computational costs and a large number of iterations. On the other hand, local search algorithms focus on exploring the immediate neighborhood of a function to find the best incremental improvement. They can be more efficient for smaller problems but may stop at local optima.

This chapter closes the book by discussing two popular local search algorithms: amoeba optimization and simulated annealing. The amoeba optimization is remarkable for its simplicity and effectiveness where geometric operations on a simplex are used to explore the solution space. The simulated annealing optimization, on the other hand, seeks inspiration from the metallurgic annealing process where the exploration-exploitation trade-off is addressed via a cooling schedule that allows progressively refined solutions.

13.1 AMOEBA OPTIMIZATION

The Amoeba method [199], also known as the Nelder[a]-Mead[b] algorithm, is a powerful optimization technique used in various fields, from engineering to finance. However, before we dive into the details of how it works, we need to review a few mathematical concepts. One of these concepts is a polytope, which plays a crucial role in the implementation of the algorithm.

A polytope P is the convex hull[c] of a finite set of vectors $\{\mathbf{v}_1, \mathbf{v}_2, \ldots, \mathbf{v}_k\} \in \mathbb{R}^d$, if and only if, no point in this set can be expressed as a linear combination of the other vectors (affinely independent), and $P = \{x \in \mathbb{R}^d | x = \sum_i^k \lambda_i \mathbf{v}_i\}$, where $\lambda_i \geq 0$ and $\sum_i^k \lambda_i = 1$. The dimension of a d-polytope is the maximum number of affinely independent points $d+1$ of the polytope subtracted by 1. Faces of a $(d+1)$-polytope can be understood as sets of points in P that compose a d-polytope. Furthermore, those faces may have $(d-1)$-polytopes in common. For example, a 2-polytope is a polygon, and a 3-polytope is a polyhedron that has flat polygonal faces.

The smallest possible polytope with non-zero volume in n-dimensional space is called *simplex*. For example, a 0-dimensional simplex is a point, a 1-dimensional simplex is a line segment, and a 2-dimensional simplex is a triangle.

The amoeba method iteratively progresses a simplex in the direction of better function values, with the hope of finding the extremum of the function within a region of interest. At each iteration, the algorithm evaluates the function at a new point that is generated by different operations: reflection, expansion, or contraction of the

---

[a]John Ashworth Nelder (1924–2010) British statistician, a fellow of the Royal Society and winner of the silver and gold Guy medals in 1977 and 2005, respectively.
[b]Roger Mead (1938–2015) English statistician.
[c]The convex hull of a finite set X is the smallest convex set containing X.

DOI: 10.1201/9781003350101-13

simplex. Depending on the success of the procedure, the simplex is updated, and the process repeats until a stopping criterion is met. One of the strengths of the method is that it does not rely on gradients and does not require the objective function to be differentiable, which makes it applicable to a wide range of problems.

The amoeba algorithm works by ordering the points of the simplex according to the objective function, then finding the centroid of the simplex, and finally applying transformations to the simplex to try to find a better configuration.

Figure 13.1 A bidimensional simplex with their points ordered according to the objective function

For an N-dimensional problem, the algorithm begins by defining an initial simplex composed of a set of points $\{x_1, \ldots, x_{N+1}\}$. These points are ordered according to the value of the objective function: $f(x_1) \leq f(x_2), \leq \ldots \leq f(x_{N+1})$.

Once the points are ordered, the centroid x_0 of points $\{x_1, \ldots, x_N\}$ can be computed using a measure[d] G_P of the set. For a finite set of points, the centroid can be estimated as:

$$x_0 = \frac{\int x \, dG_P(x)}{\int dG_P(x)} = \frac{1}{k} \sum_i^N x_i. \tag{13.1}$$

Next, simplex transformations are tested. A reflection of the worst point for the centroid is computed as:

$$x_r = x_0 + \alpha(x_0 - x_{N+1}), \tag{13.2}$$

where $\alpha > 0$ is a hyperparameter (usually $\alpha \approx 1$).

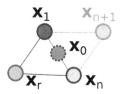

Figure 13.2 The operation of reflection on the simplex

If the reflected point is a better solution, then we can test if an expansion results in an even better situation:

$$x_e = x_0 + \gamma(x_r - x_0), \tag{13.3}$$

where $\gamma > 1$ is another hyperparameter (usually $\gamma \approx 2$).

---

[d] $G_P(x) = 1$ if $x \in P$ and 0 otherwise.

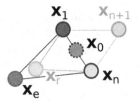

Figure 13.3   The operation of expansion on the simplex

If reflection and expansion did not result in a better candidate, we can try to contract the simplex. A contraction can be outside the simplex toward the reflected point:

$$\mathbf{x}_c = \mathbf{x}_0 + \rho(\mathbf{x}_r - \mathbf{x}_0), \tag{13.4}$$

where $0 < \rho \leq 0.5$ is another hyperparameter (usually $\rho \approx 0.5$).

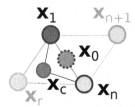

Figure 13.4   The operation of contraction outside on the simplex

Another possibility is to make a contraction inside the simplex toward the worst point:

$$\mathbf{x}_c = \mathbf{x}_0 + \rho(\mathbf{x}_{N+1} - \mathbf{x}_0). \tag{13.5}$$

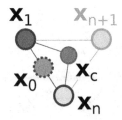

Figure 13.5   The operation of contraction inside the simplex

Finally, if the simplex transformations fail, the simplex is probably surrounding the best solution and it can be shrunk toward the best point:

$$\mathbf{x}_i = \mathbf{x}_1 + \beta(\mathbf{x}_i - \mathbf{x}_1), \ \forall(N+1 \geq i > 1), \tag{13.6}$$

where $0 < \beta < 1$ is another hyperparameter (usually $\beta \approx 0.5$).

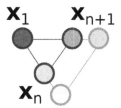

Figure 13.6   The operation of shrinking the simplex toward the best point

Given these operations, the algorithm reads:

Algorithm 29: Amoeba optimization

---

repeat

    Sort points of simplex according to an objective function f;

    Compute the centroid $\mathbf{x}_0$ of $\{\mathbf{x}_1, \ldots, \mathbf{x}_N\}$;

    Compute reflected point $\mathbf{x}_r = \mathbf{x}_0 + \alpha(\mathbf{x}_0 - \mathbf{x}_{N+1})$;

    if $f(\mathbf{x}_r) < f(\mathbf{x}_1)$ **then**

        Compute expanded point $\mathbf{x}_e = \mathbf{x}_0 + \gamma(\mathbf{x}_r - \mathbf{x}_0)$;

        if $f(\mathbf{x}_r) \leq f(\mathbf{x}_e)$ **then**

            Substitute $\mathbf{x}_{N+1} \leftarrow \mathbf{x}_r$;

        else

            Substitute $\mathbf{x}_{N+1} \leftarrow \mathbf{x}_e$;

        end

    else

        if $f(\mathbf{x}_1) \leq f(\mathbf{x}_r) < f(\mathbf{x}_N)$ **then**

            Substitute $\mathbf{x}_{N+1} \leftarrow \mathbf{x}_r$;

        else

            if $f(\mathbf{x}_r) < f(\mathbf{x}_{N+1})$ **then**

                Compute outside contraction: $\mathbf{x}_c = \mathbf{x}_0 + \rho(\mathbf{x}_r - \mathbf{x}_0)$;

            else

                Compute inside contraction: $\mathbf{x}_c = \mathbf{x}_0 + \rho(\mathbf{x}_{N+1} - \mathbf{x}_0)$;

            end

            if $f(\mathbf{x}_c) < f(\mathbf{x}_{N+1})$ **then**

                Substitute $\mathbf{x}_{N+1} \leftarrow \mathbf{x}_c$;

            else

                Shrink simplex: $\mathbf{x}_{i \neq 1} = \mathbf{x}_1 + \beta(\mathbf{x}_i - \mathbf{x}_1)$;

            end

        end

    end

end

until *stop criterium has been reached*;

Figure 13.7 shows the result of the amoeba optimization applied to the Rosenbrock function.

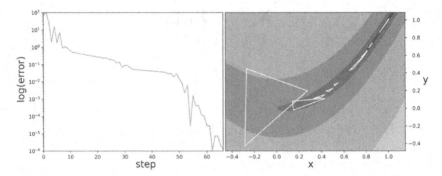

Figure 13.7 Left: The logarithm of the error as a function of the simulation step for the amoeba algorithm, and Right: A few snapshots of bidimensional simplexes approaching the minimum of the Rosenbrock function at $(1,1)$

13.2 SIMMULATED ANNEALING

As described in Section 2.4, a Markov chain describes a set of memoryless random events, which in a Monte Carlo simulation correspond to candidate solutions for a specific problem. The likelihood of finding the system in a particular state ϕ is given by the Boltzmann factor:

$$P(\phi) = \frac{1}{Z} e^{-\beta H(\phi)}, \tag{13.7}$$

where $\beta = (k_B T)^{-1}$ is the inverse temperature, $k_B = 8.617333262 \times 10^{-5}$ eV/K is the Boltzmann constant, H is the Hamiltonian of the system, and Z is the partition function, a normalizing constant that ensures that the probabilities of all candidate solutions add up to 1.

Thermal equilibrium corresponds to an irreducible, aperiodic, positive recurrent, and reversible Markov chain that satisfies the detailed balance principle for two states $\mathbf{x}$ and $\mathbf{y}$:

$$P(\mathbf{x}|\mathbf{y})\pi(\mathbf{y}) = P(\mathbf{y}|\mathbf{x})\pi(\mathbf{x}), \tag{13.8}$$

where π is a stationary probability.

The transition probabilities can be written as:

$$P(\mathbf{x}|\mathbf{y}) = Q(\mathbf{x}|\mathbf{y})A(\mathbf{x},\mathbf{y}), \tag{13.9}$$

where $Q(\mathbf{x}|\mathbf{y})$ is the probability of choosing state $\mathbf{x}$ if the chain is at state $\mathbf{y}$, and $A(\mathbf{x},\mathbf{y})$ is the acceptance ratio. To satisfy the detailed balance equation, we must have:

$$A(\mathbf{x},\mathbf{y})Q(\mathbf{x}|\mathbf{y})\pi(\mathbf{y}) = A(\mathbf{y},\mathbf{x})Q(\mathbf{y}|\mathbf{x})\pi(\mathbf{x}). \tag{13.10}$$

The acceptance ratio is a probability, and as such, it must be smaller or equal to one. Consequently, if $Q(\mathbf{y}|\mathbf{x})\pi(\mathbf{x}) > Q(\mathbf{x}|\mathbf{y})\pi(\mathbf{y})$ then $A(\mathbf{x},\mathbf{y}) = 1$ and $A(\mathbf{y},\mathbf{x}) = Q(\mathbf{x}|\mathbf{y})\pi(\mathbf{y})/Q(\mathbf{y}|\mathbf{x})\pi(\mathbf{x})$. Switching $\mathbf{x}$ and $\mathbf{y}$ gives the same result. Therefore:

$$A(\mathbf{x},\mathbf{y}) = \min\left(1, \frac{Q(\mathbf{x}|\mathbf{y})\pi(\mathbf{y})}{Q(\mathbf{y}|\mathbf{x})\pi(\mathbf{x})}\right). \tag{13.11}$$

We can select the probability of picking a state in a symmetrical manner such that $\mathbf{y} = \mathbf{x} + \eta$, where η is drawn from a symmetric probability density g centered around 0. Therefore:

$$Q(\mathbf{x}|\mathbf{y}) = g(\eta),$$
$$Q(\mathbf{y}|\mathbf{x}) = g(-\eta) = g(\eta). \tag{13.12}$$

Consequently, the acceptance rate is given by:

$$A(\mathbf{x}, \mathbf{y}) = \min\left(1, e^{-\beta \Delta_\varepsilon}\right), \tag{13.13}$$

where $\Delta_\varepsilon = H(\mathbf{x}) - H(\mathbf{y})$.

This leads to the Metropolis[e]-Hastings[f] algorithm [200, 201] used for obtaining a sequence of random samples that converge to the stationary distribution:

Algorithm 30: Metropolis-Hastings

▷ Choose initial state $\mathbf{x}_0$;
▷ Set $t = 0$;
repeat
   | ▷ Generate a candidate state $\mathbf{y} = \mathbf{x} + \eta$;
   | ▷ Find the acceptance rate: $A(\mathbf{y}, \mathbf{x}) = \min\left(1, e^{-\beta\left(\varepsilon_\mathbf{y} - \varepsilon_{\mathbf{x}_t}\right)}\right)$;
   | ▷ Draw a random number r from a uniform distribution $u_{[0,1]}$;
   | **if** $r \leq A(\mathbf{y}, \mathbf{x}_t)$ **then**
   | | ▷ Accept the candidate state $\mathbf{x}_{t+1} = \mathbf{y}$;
   | **else**
   | | ▷ Reject the candidate state and stay at the same state $\mathbf{x}_{t+1} = \mathbf{x}_t$;
   | **end**
   | ▷ Increment time $t = t + 1$;
until *stop criterium has been reached*;

Simulated annealing [202] can be viewed as a special case of the Metropolis-Hastings algorithm where the temperature controls the exploration-exploitation tradeoff. Every time an equilibrium is reached, the temperature is reduced and the whole process is repeated. Similar to a metallurgical annealing process, the algorithm approaches the global minimum of an objective function.

13.3 CLOSING REMARKS

In this comprehensive journey through machine learning, we covered a wide range of topics, starting with the mathematical foundations that form the fundamental building blocks in this field. We then delved into unsupervised and supervised learning, and, building upon these fundamentals, we explored the field of neural-inspired

---

[e]Nicholas Constantine Metropolis (1915–1999) American physicist.
[f]Wilfred Keith Hastings (1930–2016) Canadian statistician.

learning. There, we studied a variety of architectures, such as multi-layer feedforward networks, convolutional neural networks, and Boltzmann machines. We then embarked on the field of reinforcement learning that opens possibilities for autonomous decision-making and game-playing agents. Lastly, we explored optimization techniques, including population-based metaheuristics and local search methods, which are used to find solutions in complex problem spaces.

As we close this book, we have only scratched the surface of the vast field of machine learning. This is a very active and constantly progressing field, where many impactful contributions are published on a daily basis. Machine learning is revolutionizing many industries, and its potential for scientific discovery is formidable. We hope the material covered in this book has provided a solid foundation for a deeper study of the subject and hope it can be used to move the field forward.

A Sufficient Statistic

Given a sample $X = \{X_1, \ldots, X_N\}$ of some population Ω, a statistic $T(X)$ is sufficient for some model parameter θ if the conditional probability of X given $T(X)$ is independent of θ.

Mathematially, let $f_\theta(x)$ be the joint probability density function of X, then:

$$
\begin{aligned}
f_\theta(x) &= P_\theta(X = x, T(X) = T(x)) \\
&= P(X = x | T(X) = T(x)) P_\theta(T(X) = T(x)) \quad\quad\quad\quad \text{(A.1)} \\
&= g_\theta(T(x)) h(x),
\end{aligned}
$$

where $g_\theta(z) = P_\theta(T(X) = z)$ is a function that depends on the parameter θ, and $h(x) = P(X = x | T(X) = t)$ is a function that does not depend on θ.

If we now accept that $f_\theta(x)$ has the form derived in Eq. A.1, then we can write:

$$
\begin{aligned}
P_\theta(X = x | T(X) = T(x)) &= \frac{P_\theta(X = x, T(X) = T(x))}{P_\theta(T(X) = T(x))} \\
&= \frac{g_\theta(T(x)) h(x)}{\sum_{y \in A_x} f_\theta(y)}, \quad\quad\quad\quad \text{(A.2)}
\end{aligned}
$$

where $A_z = \{k : T(k) = T(z)\}$ is a partition of T. Consequently:

$$
\begin{aligned}
P_\theta(X = x | T(X) = T(x)) &= \frac{g_\theta(T(x)) h(x)}{\sum_{y \in A_x} g_\theta(T(y)) h(y)} \\
&= \frac{g_\theta(T(x)) h(x)}{g_\theta(T(x)) \sum_{y \in A_x} h(y)} \quad\quad\quad\quad \text{(A.3)} \\
&= \frac{h(x)}{\sum_{y \in A_x} h(y)},
\end{aligned}
$$

which is independent of θ. This is known as Fisher's factorization theorem [203].

As an example, consider the Poisson distribution:

$$
\begin{aligned}
f_\theta(\mathbf{x}) = P(\mathbf{x}) &= \prod_{i=1}^{N} \frac{\theta^{x_i} e^{-\theta}}{x_i!} \\
&= \left(\theta^{\sum_{i=1}^{N} x_i} e^{-N\theta} \right) \frac{1}{\prod_{i=1}^{N} x_i!}. \quad\quad\quad\quad \text{(A.4)}
\end{aligned}
$$

Here, we can identify $g_\theta(\mathbf{x}) = \theta^{\sum_{i=1}^{N} x_i} e^{-N\theta}$ and $h(\mathbf{x}) = \frac{1}{\prod_{i=1}^{N} x_i!}$. Consequently $T(\mathbf{x}) = \sum_{i=1}^{N} x_i$ is a sufficient statistic for θ.

B Graphs

A graph can be represented by many data structures such as adjacency lists, adjacency matrices, and incidence matrices. Here, we will represent a graph using a list of nodes and a list of edges inside a mutable structure. Also, inside this structure there will be a counter for the number of edges and vertices:

```
mutable struct graph
    LE :: Vector
    D :: Dict

    nnodes :: Integer
    nedges :: Integer
end
```

Note that the list of edges *LE* is a vector, while the list of nodes *D* is a dictionary.

To work with graphs, we need operations for inserting and deleting edges and vertices. These operations are simple to implement using the chosen data structure. For inserting a node, for example, we just expand the dictionary:

```
function insertNode(G::graph,n)
    G.nnodes = G.nnodes + 1
    merge!(G.D,Dict(n=>G.nnodes))
end
```

For inserting an edge, we just need to append it to the list of edges:

```
function insertEdge(G::graph,n1,n2)
    if (haskey(G.D,n1) && haskey(G.D,n2))
        push!(G.LE,[G.D[n1], G.D[n2]])
        G.nedges = G.nedges + 1
    end
end
```

DOI: 10.1201/9781003350101-B

Note that we use the dictionary to store the edges, and the function haskey guarantees that the two nodes exist.

Deleting an edge consists of finding its position in the list of edges, and then deleting this entry:

```
function deleteEdge(G::graph,n1,n2)
    p = findEdgePosition(G,G.D[n1],G.D[n2])
    deleteat!(G.LE,p)
    G.nedges = G.nedges - 1
end
```

Function findEdgePosition, as its name implies, sweeps the list of edges until the edge of interest is found.

A little care must be taken when deleting a node since all edges connected to it also have to be removed:

```
function deleteNode(G::graph,n)
    if (haskey(G.D,n))
        # First delete edges
        i = 1
        while(i <= length(G.LE))
            if (G.LE[i][1] == G.D[n] —— G.LE[i][2] == G.D[n])
                deleteat!(G.LE,i)
                G.nedges = G.nedges - 1
            else
                i = i + 1
            end
        end
        # Then, delete node
        delete!(G.D,n)
        G.nnodes = G.nnodes - 1
    end
end
```

C Sequential Minimization Optimization

For the problem [89]:

$$\begin{cases} \mathcal{L} = 1/2 \sum_{i,j} \lambda_i \lambda_j d_i d_j K_{ij} - \sum_i \lambda_i \\ \text{s.t. } \sum_i \lambda_i d_i = 0 \\ 0 \le \lambda_i \le C, \end{cases} \tag{C.1}$$

the primal feasibility of the Karush-Kuhn-Tucker[a] [88] conditions are:

$$\lambda_i \left[d_i (\mathbf{n} \cdot \mathbf{x}_i + b) - 1 + \xi_i \right] = 0$$
$$\mu_i \xi_i = 0. \tag{C.2}$$

Considering that $\lambda_i = C - \mu_i$, the last condition becomes $\xi_i(C - \lambda_i) = 0$. Therefore, if $\lambda_i = 0$, then $\mu_i = C$ and ξ_i must be 0. Therefore, the feasibility condition (Eq. 7.46) becomes:

$$d_i (\mathbf{n} \cdot \mathbf{x}_i + b) \ge 1. \tag{C.3}$$

If $\lambda_i = C$ then:

$$d_i (\mathbf{n} \cdot \mathbf{x}_i + b) - 1 = -\xi_i \le 0, \tag{C.4}$$

because $\xi_i \ge 0$. Finally, if $0 \le \lambda_i \le C$, then ξ_i must be 0, and:

$$d_i (\mathbf{n} \cdot \mathbf{x}_i + b) = 1. \tag{C.5}$$

The minimization problem can now be solved in pairs of Lagrange multipliers in a hill climbing manner. One picks one multiplier that violates the KKT conditions and optimizes the pair. The procedure is repeated until convergence is obtained.

To find the optimization of the pair, we write:

$$\mathcal{L} = 1/2 \sum_{i=1}^{2} \sum_{j=1}^{2} \lambda_i \lambda_j d_i d_j K_{ij} + 1/2 \sum_{i=1}^{2} \sum_{j=3}^{N} \lambda_i \lambda_j d_i d_j K_{ij} + 1/2 \sum_{i=3}^{N} \sum_{j=1}^{2} \lambda_i \lambda_j d_i d_j K_{ij}$$
$$+ \sum_{i=3}^{N} \sum_{j=3}^{N} \lambda_i \lambda_j d_i d_j K_{ij} - \lambda_1 - \lambda_2 - \sum_{i=3}^{N} \lambda_i. \tag{C.6}$$

---

[a]Harold William Kuhn (1925–2014) American mathematician, winner of the John von Neumann prize in 1980 together with Albert William Tucker (1905–1995) Canadian mathematician adviser of John Forbes Nash, winner of the John von Neumann prize in 1980, and William Karush (1917–1997) American mathematician.

DOI: 10.1201/9781003350101-C

To make calculations simpler, let $s = d_1 d_2$ and $v_i = \sum_{j=3}^{N} d_j \lambda_j K_{ij}$. The above equation can then be rewritten as:

$$\mathcal{L} = \tfrac{1}{2}K_{11}\lambda_1^2 + \tfrac{1}{2}K_{22}\lambda_2^2 + sK_{12}\lambda_1\lambda_2 + \lambda_1 d_1 v_1 + \lambda_2 d_2 v_2 - \lambda_1 - \lambda_2 + \mathcal{L}_0, \quad \text{(C.7)}$$

where

$$\mathcal{L}_0 = \tfrac{1}{2}\sum_{i=3}^{N}\sum_{j=3}^{N}\lambda_1\lambda_2 d_i d_j K_{ij} - \sum_{i=3}^{N}\lambda_i.$$

From the second line of Eq. C.1, we have:

$$\lambda_1 d_1 + \lambda_2 d_2 + \sum_{i=3}^{N}\lambda_i d_i = 0 \rightarrow \lambda_1 d_1 + \lambda_2 d_2 = \gamma. \quad \text{(C.8)}$$

If $d_1 = d_2$, then $\lambda_1 + \lambda_2 = \gamma$. However, if $d_1 \neq d_2$, then $\lambda_1 - \lambda_2 = \gamma$. These two situations can be combined into a single equation by writing $\lambda_1 + s\lambda_2 = \gamma \rightarrow \lambda_1 = \gamma - s\lambda_2$. Now, applying this into Eq. C.7:

$$\mathcal{L} = \tfrac{1}{2}K_{11}(\gamma - s\lambda_2)^2 + \tfrac{1}{2}K_{22}\lambda_2^2 + sK_{12}(\gamma - s\lambda_2)\lambda_2 + (\gamma - s\lambda_2)d_1 v_1 + \lambda_2 d_2 v_2$$
$$- \gamma + s\lambda_2 - \lambda_2 + \mathcal{L}_0.$$
$$\text{(C.9)}$$

Its first derivative with respect to the Lagrange multiplier is:

$$\frac{\partial \mathcal{L}}{\partial \lambda_2} = -sK_{11}(\gamma - s\lambda_2) + K_{22}\lambda_2 + sK_{12}\gamma - 2s^2 K_{12}\lambda_2 + d_2(v_2 - v_1) + s - 1, \quad \text{(C.10)}$$

where we used the trick: $sd_1 = d_1^2 d_2 = d_2$.

Its second derivative, which we will call η, becomes:

$$\frac{\partial^2 \mathcal{L}}{\partial \lambda_2^2} = K_{11} + K_{22} - 2K_{12} = \eta, \quad \text{(C.11)}$$

since $s^2 = 1$.

Using Eq. 7.43 in the argument of Eq. 7.37, we obtain the output of the SVM for an input $\mathbf{x}_1$:

$$u_1 = \sum_i \lambda_i d_i K_{1i} + b = \lambda_1 d_1 K_{11} + \lambda_2 d_2 K_{12} + v_1 + b$$
$$\text{(C.12)}$$
$$v_1 = u_1 - \lambda_1 d_1 K_{11} - \lambda_2 d_2 K_{12} - b.$$

Doing the same for the other input, we get:

$$v_2 = u_2 - \lambda_1 d_1 K_{21} - \lambda_2 d_2 K_{22} - b. \quad \text{(C.13)}$$

Now, the maximum of the Lagrangean function is obtained when its derivative (Eq. C.10) is zero. Using, the previous two results, we get:

$$\lambda_2(K_{11} + K_{22} - 2K_{12}) = sK_{11}\gamma - sK_{12}\gamma - d_2(v_2 - v_1) - s + 1$$
$$\eta\lambda_2^{\text{new}} = s\gamma(K_{11} - K_{12}) - d_2(v_2 - v_1) - d_1 d_2 + d_2 d_2 \quad \text{(C.14)}$$
$$\lambda_2^{\text{new}} = \lambda_2^{\text{old}} - d_2\frac{e_2 - e_1}{\eta},$$

where $e_i = u_i - d_i$.

This is an unconstrained solution but we must guarantee that $0 \leq \lambda_i \leq C$ by clipping. Therefore, if $s = 1$, then $\lambda_1 + \lambda_2 = \gamma$. If $\gamma > C$ then $\lambda_2^{\max} = C$ and $\lambda_2^{\min} = \gamma - C$. If $\gamma < C$, then $\lambda_2^{\min} = 0$ and $\lambda_2^{\max} = \gamma$. Now, if $s = -1$, then $\lambda_1 - \lambda_2 = \gamma$. If $\gamma > 0$, then $\lambda_2^{\min} = 0$ and $\lambda_2^{\max} = C - \gamma$. And if $\gamma < 0$, then $\lambda_2^{\min} = -\gamma$ and $\lambda_2^{\max} = C$.

With this clipped multiplier, we can find the other, given that:

$$\lambda_1^{\text{new}} + s\lambda_2^{\text{clipped}} = \lambda_1^{\text{old}} + s\lambda_2^{\text{old}}. \tag{C.15}$$

Therefore:

$$\lambda_1^{\text{new}} = \lambda_1^{\text{old}} + s\left(\lambda_2^{\text{old}} - \lambda_2^{\text{clipped}}\right). \tag{C.16}$$

D Algorithmic Differentiation

Algorithmic differentiation (AD) [204], also known as *automatic differentiation*, is a computational technique that can be used to differentiate functions. When a program employs this method, it is referred to as *differential programming*.

AD can be performed in forward and reverse modes. In the forward mode, derivatives are calculated using *dual numbers*[a]. In abstract algebra, dual numbers are defined as a commutative unital algebra[b] over the real numbers with two generators, 1 and ε, subject to the defining relation[c] $\varepsilon^2 = 0$. The elements of the dual numbers can be represented[d] as $a + b\varepsilon$, where $a, b \in \mathbb{R}$. Therefore, for any analytical function f, we have a Taylor expansion around x:

$$
\begin{aligned}
f(a+b\varepsilon) &= \sum_{n=0}^{\infty} \frac{1}{n!} f^{(n)}(x)(a-b\varepsilon-x)^n \\
&= \sum_{n=0}^{\infty} \frac{1}{n!} f^{(n)}(x) \left[(a-x)^n - n(a-x)^{n-1} b\varepsilon \right] \qquad \text{(D.1)} \\
&= f(a) + \left. \frac{d}{da} f(a) \right|_{a=x} b\varepsilon.
\end{aligned}
$$

Consequently, in a forward pass, both the value of the function and its derivative are computed simultaneouly. For example, in Julia we may write:

```
using DualNumbers

x1 = Dual(2,1)
y2 = Dual(3,0)
x1 = Dual(2,0)
y2 = Dual(3,1)

fx = x1^2+x1*y1
fy = x2^2+x2*y2
```

[a] Introduced by William Kingdon Clifford (1854–1879) English mathematician.
[b] A vector space A over a field K equipped with a bilinear product $A \times A \to A$ that is associative, distributive, and has an identity element with respect to multiplication.
[c] Dual numbers are 2-dimensional *hypercomplex numbers* together with *double numbers* where $\varepsilon^2 = 1$, and *complex numbers* where $\varepsilon^2 = -1$.
[d] They can also be represented by the matrix $\begin{bmatrix} a & b \\ 0 & 0 \end{bmatrix}$.

We obtain $fx = 10 + 7\varepsilon$ and $fy = 10 + 2\varepsilon$. It is important to note that, to obtain the derivative with respect to a parameter, a separate calculation must be performed for each parameter of interest.

In the reverse mode, the sequence of elementary operations, also known as the *trace* or *Wengert list*, is used to construct a direct acyclic graph (DAG). The partial derivatives are calculated by traversing the graph in reverse topological order and, consequently, by applying the chain rule. For example, for the same function $f(x,y) = x^2 + xy$, we obtain the DAG shown in Fig. D.1. In this DAG, we use some intermediate variables $u = x^2$ and $v = xy$, so that $f = u + v$.

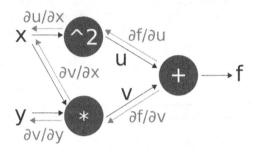

Figure D.1   Direct acyclic graph for the function $f(x,y) = x^2 + xy$.

To obtain the derivative of f with respect to x, for example, we traverse the DAG in reverse topological order, and obtain:

$$\begin{aligned}
\frac{df}{dx} &= \frac{\partial f}{\partial u}\frac{\partial u}{\partial x} + \frac{\partial f}{\partial v}\frac{\partial v}{\partial x} \\
&= 1 \cdot 2x + 1 \cdot y \\
&= 2x + y.
\end{aligned} \tag{D.2}$$

For the input $x,y = 2,3$, we get $df/dx = 7$ as before. The reverse mode consists of a forward phase, where the nodes are evaluated with the desired input, and a backward phase, where the partial derivatives are computed. The attentive reader must have realized that this is exactly the backpropagation used in neural networks. Furthermore, while the chain rule dates back to Leibiniz[e], it has gained renewed appreciation in the context of neural networks.

One of the key advantages of the reverse mode over the forward mode is that it allows for efficient reuse of partial derivatives, enabling the backward traversal of the computation graph just once. However, for $f : \mathbb{R}^n \to \mathbb{R}^m$, the forward mode is computationally more efficient if $n \ll m$.

---

[e]Gottfried Wilhelm von Leibniz (1646–1716) German polymath, partially advised by Christiaan Huygens, and epistolary adviser of Jacob Bernoulli.

E Batch Normalizing Transform

Consider a mini-batch of size N_{MB}. For each mini-batch b, a vector $\mathbf{x}^{(b)} = \left(x_1^{(b)}, x_2^{(b)}, \ldots, x_N^{(b)} \right)$ is input to the neural network. The mean and average for each mini-batch are given by:

$$\mu_i = \frac{1}{N_{MB}} \sum_{b=1}^{N_{MB}} x_i^{(b)} \rightarrow \mu = \frac{1}{N_{MB}} \sum_{b=1}^{N_{MB}} \mathbf{x}^{(b)}$$

$$\sigma_i^2 = \frac{1}{N_{MB}} \sum_{b=1}^{N_{MB}} (x_i^{(b)} - \mu_i)^2 \rightarrow \sigma^2 = \frac{1}{N_{MB}} \sum_{b=1}^{N_{MB}} (\mathbf{x}^{(b)} - \mu)^{\circ 2}.$$

(E.1)

The input is then normalized, shifted and rescaled, as:

$$\hat{x}_i^{(b)} = \frac{x_i^{(b)} - \mu_i}{\sqrt{\sigma_i^2 + \varepsilon^2}}$$

$$\hat{\mathbf{x}}^{(b)} = \left(\mathbf{x}^{(b)} - \mu \right) \oslash \left(\sigma^2 + \varepsilon \right)^{\circ 1/2}.$$

(E.2)

Finally, the layer makes a mapping:

$$y_i^{(b)} = \gamma_i \hat{x}_i^{(b)} + \beta_i$$

$$\mathbf{y}^{(b)} = \gamma \odot \hat{\mathbf{x}}^{(b)} + \beta,$$

(E.3)

where γ and β are parameters to be learned.

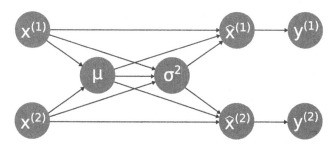

Figure E.1 Batch transform scheme for a mini batch of size 2. Observe that the speheres are not neurons, but mini matches

DOI: 10.1201/9781003350101-E

For the backpropagation, according to Fig. E.1, we have:

$$\frac{\partial \mathbf{y}^{(b)}}{\partial \beta} = \mathbf{I}$$

$$\frac{\partial \mathbf{y}^{(b)}}{\partial \gamma} = diag\left(\hat{\mathbf{x}}^{(b)}\right)$$

$$\frac{\partial \mathbf{y}^{(b)}}{\partial \mathbf{x}^{(c)}} = \frac{\partial \mathbf{y}^{(b)}}{\partial \hat{\mathbf{x}}^{(b)}} \frac{\partial \hat{\mathbf{x}}^{(b)}}{\partial \mathbf{x}^{(c)}} \delta_{bc} + \frac{\partial \mathbf{y}^{(b)}}{\partial \hat{\mathbf{x}}^{(b)}} \frac{\partial \hat{\mathbf{x}}^{(b)}}{\partial \sigma^2} \frac{\partial \sigma^2}{\partial \mathbf{x}^{(c)}} + \frac{\partial \mathbf{y}^{(b)}}{\partial \hat{\mathbf{x}}^{(b)}} \frac{\partial \hat{\mathbf{x}}^{(b)}}{\partial \sigma^2} \frac{\partial \sigma^2}{\partial \mu} \frac{\partial \mu}{\partial \mathbf{x}^{(c)}}$$

$$+ \frac{\partial \mathbf{y}^{(b)}}{\partial \hat{\mathbf{x}}^{(b)}} \frac{\partial \hat{\mathbf{x}}^{(b)}}{\partial \mu} \frac{\partial \mu}{\partial \mathbf{x}^{(c)}}.$$

(E.4)

The common term that appears in all components is:

$$\frac{\partial \mathbf{y}^{(b)}}{\partial \mathbf{x}^{(b)}} = diag(\gamma) = \gamma_d.$$

(E.5)

Therefore, the expression becomes:

$$\frac{\partial \mathbf{y}^{(b)}}{\partial \mathbf{x}^{(c)}} = \gamma_d \left(\frac{\partial \hat{\mathbf{x}}^{(b)}}{\partial \mathbf{x}^{(c)}} \delta_{bc} + \frac{\partial \hat{\mathbf{x}}^{(b)}}{\partial \sigma^2} \left(\frac{\partial \sigma^2}{\partial \mathbf{x}^{(c)}} + \frac{\partial \sigma^2}{\partial \mu} \frac{\partial \mu}{\partial \mathbf{x}^{(c)}} \right) + \frac{\partial \hat{\mathbf{x}}^{(b)}}{\partial \mu} \frac{\partial \mu}{\partial \mathbf{x}^{(c)}} \right).$$

(E.6)

The remaining derivatives are:

$$\frac{\partial \hat{\mathbf{x}}^{(b)}}{\partial \sigma^2} = -\frac{1}{2} \left(\mathbf{x}^{(b)} - \mu \right) \oslash (\sigma^2 + \varepsilon)^{\circ 3/2}$$

$$\frac{\partial \sigma^2}{\partial \mu} = -\frac{2}{N_{MB}} \sum_{b=1}^{N_{MB}} (\mathbf{x}^{(b)} - \mu)$$

$$\frac{\partial \hat{\mathbf{x}}^{(b)}}{\partial \mu} = -\mathbf{I} \oslash (\sigma^2 + \varepsilon)^{\circ 1/2}$$

$$\frac{\partial \mu}{\partial \mathbf{x}^{(c)}} = \frac{\mathbf{I}}{N_{MB}}$$

$$\frac{\partial \sigma^2}{\partial \mathbf{x}^{(c)}} = \frac{2}{N_{MB}} (\mathbf{x}^{(c)} - \mu)$$

$$\frac{\partial \hat{\mathbf{x}}^{(b)}}{\partial \mathbf{x}^{(b)}} = \mathbf{I} \oslash (\sigma^2 + \varepsilon)^{\circ 1/2} = -\frac{\partial \hat{\mathbf{x}}^{(b)}}{\partial \mu}.$$

(E.7)

With this results we can simplify Eq. E.6 to:

$$\frac{\partial \mathbf{y}^{(b)}}{\partial \mathbf{x}^{(c)}} = \gamma_d \left(\frac{\partial \hat{\mathbf{x}}^{(b)}}{\partial \mathbf{x}^{(c)}} \delta_{bc} + \frac{\partial \hat{\mathbf{x}}^{(b)}}{\partial \sigma^2} \left(\frac{\partial \sigma^2}{\partial \mathbf{x}^{(c)}} + \frac{1}{N_{MB}} \frac{\partial \sigma^2}{\partial \mu} \right) - \frac{1}{N_{MB}} \frac{\partial \hat{\mathbf{x}}^{(b)}}{\partial \mathbf{x}^{(b)}} \right)$$

$$\frac{\partial \mathbf{y}^{(b)}}{\partial \mathbf{x}^{(c)}} = \gamma_d \left(\left(\delta_{bc} - \frac{1}{N_{MB}} \right) \frac{\partial \hat{\mathbf{x}}^{(b)}}{\partial \mathbf{x}^{(b)}} + \frac{\partial \hat{\mathbf{x}}^{(b)}}{\partial \sigma^2} \left(\frac{\partial \sigma^2}{\partial \mathbf{x}^{(c)}} + \frac{1}{N_{MB}} \frac{\partial \sigma^2}{\partial \mu} \right) \right).$$

(E.8)

The loss that each minibatch element experiences is the average overall loss. Therefore, according to Fig. E.1:

$$\frac{\partial \langle L \rangle}{\partial \mathbf{x}^c} = \frac{1}{N_{MB}} \sum_{b=1}^{N_{MB}} \frac{\partial L_b}{\partial \mathbf{y}^{(b)}} \frac{\partial \mathbf{y}^{(b)}}{\partial \mathbf{x}^{(c)}}. \tag{E.9}$$

Note that it is necessary to propagate all samples in the mini batch up to the transform layer before proceeding forward since we must compute the mean and variance of the batch. While sequential computation can be costly, this function can be efficiently executed on a parallel processor.

F Divergence of two Gaussian Distributions

An n-dimensional Gaussian distribution with mean μ and covariance matrix Σ is given by:

$$f(\mathbf{x}) = \frac{1}{(2\pi)^n |\Sigma|^{1/2}} \exp\left\{ -\frac{1}{2}(\mathbf{x} - \mu)^\top \Sigma^{-1} (\mathbf{x} - \mu) \right\}. \tag{F.1}$$

The Kullback-Leibler divergence of two n-dimensional Gaussian distributions p and q is:

$$\begin{aligned} D_{KL}(p||q) &= \left\langle \log \left(\frac{p}{q} \right) \right\rangle_p \\ &= \left\langle \log \frac{|\Sigma_q|^{1/2}}{|\Sigma_p|^{1/2}} + \frac{1}{2}(\mathbf{x} - \mu_q)^\top \Sigma_q^{-1}(\mathbf{x} - \mu_q) \right. \\ &\qquad \left. - \frac{1}{2}(\mathbf{x} - \mu_p)^\top \Sigma_p^{-1}(\mathbf{x} - \mu_p) \right\rangle_p. \end{aligned} \tag{F.2}$$

Since the arguments are scalars, we can write them as traces, and using the clyciy permutation property of traces, we obtain:

$$\begin{aligned} D_{KL}(p||q) = \frac{1}{2} \left[\log \frac{|\Sigma_q|}{|\Sigma_p|} + \mathrm{Tr} \left\langle \Sigma_q^{-1}(\mathbf{x} - \mu_q)(\mathbf{x} - \mu_q)^\top \right\rangle_p \right. \\ \left. - \mathrm{Tr} \left\langle \Sigma_p^{-1}(\mathbf{x} - \mu_p)(\mathbf{x} - \mu_p)^\top \right\rangle_p \right]. \end{aligned} \tag{F.3}$$

The second expectation gives exactly the covariance matrix Σ_p. Therefore:

$$D_{KL}(p||q) = \frac{1}{2} \left[\log \frac{|\Sigma_q|}{|\Sigma_p|} + \mathrm{Tr} \left(\Sigma_q^{-1} \left\langle \mathbf{x}\mathbf{x}^\top - \mathbf{x}\mu_q^\top - \mu_q \mathbf{x}^\top + \mu_q \mu_q^\top \right\rangle_p \right) - n \right]. \tag{F.4}$$

The first term in the expectation produces $\Sigma_p + \mu_p \mu_p^\top$. Therefore:

$$D_{KL}(p||q) = \frac{1}{2} \left[\log \frac{|\Sigma_q|}{|\Sigma_p|} + \mathrm{Tr} \left(\Sigma_q^{-1} \left(\Sigma_p + \mu_p \mu_p^\top - \mu_p \mu_q^\top - \mu_q \mu_p^\top + \mu_q \mu_q^\top \right) \right) - n \right]. \tag{F.5}$$

Using again the cyclic permutation property of the trace, we may write:

$$D_{KL}(p||q) = \frac{1}{2} \left[\log \frac{|\Sigma_q|}{|\Sigma_p|} + \mathrm{Tr} \left(\Sigma_q^{-1} \Sigma_p \right) + \left(\mu_p - \mu_q \right)^\top \Sigma_q^{-1} \left(\mu_p - \mu_q \right) - n \right]. \tag{F.6}$$

DOI: 10.1201/9781003350101-F

If $q \sim \mathcal{N}(\mathbf{0}, \mathbf{I})$, then:

$$D_{KL}(p||q) = \frac{1}{2}\left[-\log|\Sigma| + \mathrm{Tr}(\Sigma) + \mu^\top\mu - n\right], \tag{F.7}$$

where the index p was dropped for simplicity.

The gradient of the divergence becomes:

$$\nabla D_{KL}(p||q) = \left[\mu^\top \quad \tfrac{1}{2}\left(\mathbf{1} - \mathrm{inv}(\Sigma)^\top\right)\right]. \tag{F.8}$$

The inverse of the covariance matrix, if it exists, is often called *precision matrix*.

G Continuous-time Bellman's Equation

The continuous Bellman's equation is widely used in finances. Let's take an investment interval T and divide it in T/Δ sub-intervals, each with an *interest rate* ρ. Therefore, the present value for the investment in this period is discounted by:

$$v(T) = \frac{1}{(1+\rho\Delta)^{T/\Delta}}. \tag{G.1}$$

If we take the interval Δ to be infinitesimal, we get:

$$\begin{aligned}
v(T) &= \lim_{\Delta \to 0} (1+\rho\Delta)^{-T/\Delta} \\
&= \lim_{\Delta \to 0} \exp\left\{-\frac{T}{\Delta}\log(1+\rho\Delta)\right\}, \text{ L'Hopital:} \\
&= \lim_{\Delta \to 0} \exp\left\{-T\frac{\rho}{1+\rho\Delta}\right\} \to e^{-\rho T}.
\end{aligned} \tag{G.2}$$

Therefore, each future reward is discounted according to this expression, and Eq. 10.5, in the continuous-time case, becomes:

$$R(t) = \int_t^\infty e^{-\rho(s-t)} r(s)ds. \tag{G.3}$$

For a Markov chain, the next state is given by a state equation:

$$\dot{x} = g(x,u), \tag{G.4}$$

where u is some control, and $x(0)$ is known. Hence, the maximality problem described by Eq. 10.12 becomes:

$$V^*(x,t) = \max_u \int_t^\infty e^{-\rho(s-t)} r(x(s),u(s))ds, \ t \geq 0. \tag{G.5}$$

This can be decomposed in two parts:

$$\begin{aligned}
V^*(x,t) &= \max_u \left[\int_t^{t+\Delta_t} e^{-\rho(s-t)} r(x(s),u(s))ds + \int_{t+\Delta_t}^\infty e^{-\rho(s-t)} r(x(s),u(s))ds\right] \\
&= \max_u \left[\int_t^{t+\Delta_t} e^{-\rho(s-t)} r(x(s),u(s))ds + V^*(x+g(x,t)\Delta_t, t+\Delta_t)e^{-\rho\Delta_t}\right].
\end{aligned} \tag{G.6}$$

DOI: 10.1201/9781003350101-G

The first integral can be approximated by a Riemann[a] sum, while the other can be approximated by a Taylor[b] expansion around t:

$$V^*(x,t) \approx \max_u [r(x(t),u(t))\Delta_t +$$

$$+ \left(V^*(x,t) + g(x,t)\frac{\partial}{\partial x}V^*(x,t) + \frac{\partial}{\partial t}V^*(x,t) \right)(1 - \rho\Delta_t)]. \tag{G.7}$$

After some algebra and taking the limit $\Delta_t \to 0$:

$$\rho V^*(x,t) \approx \max_u \left[r(x,u) + g(x,t)\frac{\partial}{\partial x}V^*(x,t) + \frac{\partial}{\partial t}V^*(x,t) \right]$$

$$\approx \max_u \left[r(x,u) + \frac{d}{dt}V^*(x,t) \right] \tag{G.8}$$

$$\rho V^*(x,t)dt \approx \max_u [r(x,u)dt + dV^*(x,t)].$$

If the value-state function can be written as an Ito's process:

$$dV = a(x,t)dt + b(x,t)dW, \tag{G.9}$$

then, we can apply Ito's lemma [24]:

$$dV^* \approx \left(\frac{\partial V^*}{\partial t} + a\frac{\partial V^*}{\partial x} + \frac{1}{2}b^2\frac{\partial V^*}{\partial x^2} \right)dt + b\frac{\partial V^*}{\partial x}dW. \tag{G.10}$$

to Eq. G.8 and obtain:

$$\rho V^*(x,t)dt \approx \max_u \left[r(x,u)dt + \left(\frac{\partial V^*}{\partial t} + a\frac{\partial V^*}{\partial x} + \frac{1}{2}b^2\frac{\partial V^*}{\partial x^2} \right)dt \right], \tag{G.11}$$

where the Brownian motion was removed since its expected value is zero.

---

[a]Georg Friedrich Bernhard Riemann (1826–1866) German mathematician, advisee of Carl Friedrich Gauss.

[b]Brook Taylor (1685–1731) English mathematician.

H Conjugate Gradient

Two vectors $\mathbf{u}$ and $\mathbf{v}$ are said to be conjugate with respect to a positive-definite[a] matrix $\mathbf{A}$ if $\mathbf{u}^\top \mathbf{A} \mathbf{v} = 0$. Now, consider that we have a base of conjugate vectors $\mathbf{P} = \{\mathbf{p}_1, \ldots, \mathbf{p}_N\}$ with respect to $\mathbf{A}$, such that $\mathbf{p}_i^\top \mathbf{A} \mathbf{p}_j = 0$, $\forall i \neq j$, and a system of linear equations $\mathbf{A}\mathbf{x} = \mathbf{b}$. The solution $\mathbf{x}$ to this matrix equation can be written as a linear combination of the elements of the basis: $\mathbf{x} = \sum_{i=1}^{N} \alpha_i \mathbf{p}_i$. Under these conditions, the linear system can be written as:

$$\sum_{i=1}^{N} \alpha_i \mathbf{p}_k^\top \mathbf{A} \mathbf{p}_i = \mathbf{p}_k^\top \mathbf{b} \rightarrow \alpha_k = \frac{\mathbf{p}_k^\top \mathbf{b}}{\mathbf{p}_k^\top \mathbf{A} \mathbf{p}_k}. \tag{H.1}$$

This problem would be solved if we had the orthogonal basis. Nonetheless, it is possible to proceed by recasting the linear system as a minimization of the quadratic function:

$$f(\mathbf{x}) = \frac{1}{2} \mathbf{x}^\top \mathbf{A} \mathbf{x} - \mathbf{x}^\top \mathbf{b} \rightarrow \nabla f(\mathbf{x}) = \mathbf{A}\mathbf{x} - \mathbf{b}. \tag{H.2}$$

Let's take an initial guess $\mathbf{x} = \mathbf{x}_0$. For this, we obtain a residual $r_0 = \mathbf{b} - \mathbf{A}\mathbf{x}_0$, which is exactly the negative of the gradient. We can take the first element of the basis to be exactly the first residual and then progressively produce orthogonal vectors using a process such as Gram[b]-Schmidt[c] orthonormalization [3]:

$$\mathbf{p}_k = \mathbf{r}_k - \sum_{i<k} \frac{\mathbf{p}_i^\top \mathbf{A} \mathbf{r}_k}{\mathbf{p}_i^\top \mathbf{A} \mathbf{p}_i} \mathbf{p}_i. \tag{H.3}$$

This last equation can be written as:

$$\mathbf{p}_{k+1} = \mathbf{r}_{k+1} + \beta_k \mathbf{p}_k. \tag{H.4}$$

To minimize $f(\mathbf{x})$, we should move exactly in the direction of the residual since it is the negative of the gradient:

$$\mathbf{x}_{k+1} = \mathbf{x}_k + \gamma_k \mathbf{p}_k. \tag{H.5}$$

From this, we find:

$$\begin{aligned} \mathbf{r}_{k+1} &= \mathbf{b} - \mathbf{A}\mathbf{x}_{k+1} = \mathbf{b} - \mathbf{A}\mathbf{x}_k - \gamma_k \mathbf{A}\mathbf{p}_k \\ &= \mathbf{r}_k - \gamma_k \mathbf{A}\mathbf{p}_k. \end{aligned} \tag{H.6}$$

---

[a] $\mathbf{A}$ is positive-definite if it is an $n \times n$ symmetric real matrix, such that $\mathbf{x}^\top \mathbf{A} \mathbf{x} > 0$, $\forall \mathbf{x} \in \mathbb{R}^n \setminus \{0\}$.
[b] Jørgen Pedersen Gram (1850–1916) Danish mathematician.
[c] Erhard Schmidt (1876–1959) German mathematician, advised by David Hilbert, and adviser of Eberhard Hopf, among others.

As we proceeded to find Eq. H.1, we can calculate $f(\mathbf{x}_{k+1})$, and minimize it with respect to γ_k to find:

$$\gamma_k = \frac{\mathbf{p}_k^\top (\mathbf{b} - \mathbf{A}\mathbf{x}_k)}{\mathbf{p}_k^\top \mathbf{A}\mathbf{p}_k} = \frac{\mathbf{p}_k^\top \mathbf{r}_k}{\mathbf{p}_k^\top \mathbf{A}\mathbf{p}_k} = \frac{\mathbf{r}_k^\top \mathbf{r}_k}{\mathbf{p}_k^\top \mathbf{A}\mathbf{p}_k}, \tag{H.7}$$

where we have used Eq. H.4 in the last step.

As the last step, we need to find β_k. For this, we take the transpose of Eq. H.4 and multiply it by $\mathbf{A}\mathbf{p}_k$ from the right-hand side:

$$\mathbf{p}_{k+1}^\top \mathbf{A}\mathbf{p}_k = 0 = \mathbf{r}_{k+1}^\top \mathbf{A}\mathbf{p}_k + \beta_k \mathbf{p}_k^\top \mathbf{A}\mathbf{p}_k$$

$$\therefore \beta_k = -\frac{\mathbf{r}_{k+1}^\top \mathbf{A}\mathbf{p}_k}{\mathbf{p}_k^\top \mathbf{A}\mathbf{p}_k}. \tag{H.8}$$

This can be further simplified using the result from Eq. H.4 and the fact that the residuals are orthogonal:

$$\beta_k = \frac{\mathbf{r}_{k+1}^\top \mathbf{r}_{k+1}}{\mathbf{r}_k^\top \mathbf{r}_k}. \tag{H.9}$$

With all these results, we are ready to assemble the conjugate algorithm[d] [205] illustrated below.

Algorithm 31: Conjugate gradient

---

input : Positive-definite matrix $\mathbf{A}$ and vector $\mathbf{b}$

▷ Guess $\mathbf{x}_0$, and calculate $\mathbf{p} = \mathbf{r} = \mathbf{b} - \mathbf{A}\mathbf{x}_0$;

▷ Compute $\xi = \mathbf{r}^\top \mathbf{r}$;

for $k = 1 \ldots length(\mathbf{b})$ **do**

    ▷ Compute $\mathbf{A}\mathbf{p}$ and $\gamma = \frac{\xi}{\mathbf{p}^\top \mathbf{A}\mathbf{p}}$;

    ▷ With γ, find $\mathbf{x} = \mathbf{x} + \gamma\mathbf{p}$ and $\mathbf{r} = \mathbf{r} - \gamma\mathbf{A}\mathbf{p}$;

    ▷ Compute $\xi' = \mathbf{r}^\top \mathbf{r}$;

    if ξ' *is sufficiently small* **then**

        | Exit loop;

    else

        ▷ Compute $\mathbf{p} = \mathbf{r} + \frac{\xi'}{\xi}\mathbf{p}$;

        ▷ Make $\xi = \xi'$;

    end

end

output: x

---

[d]Created by Magnus Hestenes (1906–1991) American mathematician, Eduard L. Stiefel (1909–1978) Swiss mathematician, and Cornelius Lanczos (1893–1974) Hungarian mathematician.

This produces the following Julia code:

```
r = b - A*x
p = r
ξ = r·r

for i in 1:N
    α = ξ/(p·(A*p))
    x = x + α*p
    r = r - α*A*p
    new_ξ = r·r

    if (new_ξ < 1E-10)
        break
    else
        p = r + (new_ξ/ξ)*p
        ξ = new_ξ
    end
end
```

I | Importance Sampling

Consider that you have a random variable from which it is diffult to sample. For instance, it is difficult to sample numbers greater than 5 from a $\mathcal{N}(0,1)$ distribution. Importance sampling [206] is a technique that allows us to sample from another distribution and mitigate this problem. To show this, let's begin with the definition of the expectation of a function $f(x)$ and probability measures P and Q on a measurable space $(\Omega, \mathscr{A})$. If $Q(A) = 0 \implies P(A) = 0$, $\forall A \in \mathscr{A}$, then we can perform a *change of measure*:

$$\langle f \rangle_P = \int f(x)dP = \int \frac{f(x)dP}{dQ}dQ = \left\langle \frac{fdP}{dQ} \right\rangle_Q. \tag{I.1}$$

The Radon-Nikodym derivative [24] dP/dQ is known as *likelihood ratio*, the original distribution P is known as the *nominal measure*, and Q is the *importance sampling measure*.

Often in reinforced learning, one uses a behavior policy $b(a|s)$ that explores the environment and a target policy $\pi_\theta(a|s)$, which is trained based on the explorations of the former. For this, we have to change the measure of the policy score function (Eq. 11.1):

$$
\begin{aligned}
J = \langle R(\tau) \rangle_{\text{target}} &= \int_\tau \rho(\tau)R(\tau)d\tau \\
&= \int_\tau \frac{\rho(\tau)R(\tau)}{\rho_b(\tau)}\rho_b(\tau)d\tau \\
&= \left\langle \frac{\rho R(\tau)}{\rho_b} \right\rangle_{\text{behavior}}.
\end{aligned}
\tag{I.2}
$$

The likelihood ratio can be computed with the aid of Eq. 11.4:

$$
\begin{aligned}
\frac{\rho(\tau)}{\rho_b(\tau)} &= \frac{\rho(s_0,a_0)\prod_{t=0}^{T-1}\pi_\theta(a_t|s_t)\rho(s_{t+1}|s_t,a_t)}{\rho(s_0,a_0)\prod_{t=0}^{T-1}b(a_t|s_t)\rho(s_{t+1}|s_t,a_t)} \\
&= \frac{\prod_{t=0}^{T-1}\pi_\theta(a_t|s_t)}{\prod_{t=0}^{T-1}b(a_t|s_t)} = \prod_{t=0}^{T-1}\frac{\pi_\theta(a_t|s_t)}{b(a_t|s_t)}.
\end{aligned}
\tag{I.3}
$$

From this, the score function under the new measure becomes:

$$\hat{J}(\theta) = \left\langle \left(\prod_{t=0}^{T-1}\left(\frac{\pi_\theta(a_t|s_t)}{b(a_t|s_t)} \right) \right)\left(\sum_{t=0}^{T-1}\gamma^t r_{t+1} \right) \right\rangle_{\text{behavior}}. \tag{I.4}$$

The gradient of the policy score function can be calculated from Eq. I.6:

$$
\begin{aligned}
\nabla J(\theta) &= \left\langle \frac{\nabla_\theta \rho(\tau)R(\tau)}{\rho_b(\tau)} \right\rangle_{\text{behavior}} \\
&= \left\langle \frac{\rho \nabla_\theta \log(\rho)R(\tau)}{\rho_b} \right\rangle_{\text{behavior}}.
\end{aligned}
\tag{I.5}
$$

Applying causality, as in Sec. 11.2, we get:

$$\nabla J(\theta) = \left\langle \sum_{t=0}^{T-1} \nabla_\theta \log\left(\pi_\theta(a_t|s_t)\right) \prod_{t'=0}^{t} \frac{\pi_\theta(a_t|s_t)}{b(a_t|s_t)} \right.$$
$$\left. \left(\sum_{t''=t}^{T-1} \gamma^{t''} r(s_{t''}, a_{t''}, s_{t'''}) \prod_{p=t+1}^{T-1} \frac{\pi_\theta(a_p|s_p)}{b(a_p|s_p)} \right) \right\rangle_{\text{behavior}}.$$

The second product operator gives the future probability of transitioning between states. Ignoring this term, we get:

$$\nabla J(\theta) \approx \left\langle \sum_{t=0}^{T-1} \nabla_\theta \log\left(\pi_\theta(a_t|s_t)\right) \prod_{t'=0}^{t} \frac{\pi_\theta(a_t|s_t)}{b(a_t|s_t)} \left(\sum_{t''=t}^{T-1} \gamma^{t''} r(s_{t''}, a_{t''}, s_{t'''}) \right) \right\rangle_{\text{behavior}}.$$
$$\text{(I.6)}$$

The product operator that remains can also cause stability issues. Therefore, we resort to a first-order approximation, where we substitute the products with the state-action marginal distributions $\pi_\theta(s_t, a_t) = \pi_\theta(a_t|s_t)$ and $b(s_t, a_t) = b(s_t)b(a_t|s_t)$, where $\pi(s_t)$ is the state marginal distribution. If the policies are close enough, we can write:

$$\nabla J(\theta) \approx \left\langle \sum_{t=0}^{T-1} \frac{\pi_\theta(a_t|s_t)}{b(a_t|s_t)} \nabla_\theta \log\left(\pi_\theta(a_t|s_t)\right) \left(\sum_{t''=t}^{T-1} \gamma^{t''} r(s_{t''}, a_{t''}, s_{t'''}) \right) \right\rangle_{\text{behavior}}. \quad \text{(I.7)}$$

Both Eqs. I.6 and I.7 can be used to formulate off-policy algorithms. While the first equation can cause instabilities if long periods are considered, the second equation circumvents this problem but can only be used if the policies are close to each other.

References

1. V. A. Epanechnikov. Non-parametric estimation of a multivariate probability density. *Theory Probab. Appl.*, 14(1):153–158, 1969.

2. J. D. Hunter. Matplotlib: A 2d graphics environment. *Computing in Science & Engineering*, 9(3):90–95, 2007.

3. K. Novak. *Numerical Methods for Scientific Computing*. Equal Share Press, Arlington, VA, 2 edition, 2022.

4. B. N. Cooperstein. *Advanced Linear Algebra*. CRC Press, Boca Raton, FL, 2 edition, 2015.

5. J. Bezanson, A. Edelman, S. Karpinski, and V. B. Shah. Julia: A fresh approach to numerical computing. *SIAM review*, 59(1):65–98, 2017.

6. A. Graham. *Kronecker Products & Matrix Calculus with Applications*. Dover Publications, Inc., Mineola, NY, 2018.

7. O. Knill. Probability theory and stochastic processes with applications. 2009.

8. P. Devolder, J. Janssen, and R. Manca. Basic stochastic processes. 2015.

9. G. A. Pavliotis. Stochastic processes and applications. 2014.

10. P. Billingsley. *Probability and Measure*. Wiley series in probability and mathematical statistics. Wiley, Hoboken, NJ, 3 edition, 1995.

11. N. L. Carothers. *Real analysis*. Cambridge University Press, Cambridge, UK, 2000.

12. N. Smirnov. Table for estimating the goodness of fit of empirical distributions. *Ann. Math. Stat.*, 19(2):279–281, 1948.

13. J. V. Stone. Information theory - a tutorial introduction. 2022.

14. S. Kullback and R. A. Leibler. On information and sufficiency. *Ann. Math.*, 22(1): 790086, 1951.

15. L. M. Bregman. The relaxation method of finding the common point of convex sets and its application to the solution of problems in convex programming. *USSR Comput. Math. Math. Phys.*, 7(3):200–217, 1967.

16. H. Jeffrey. *Theory of probability*. Oxford classic texts in the physics sciences. Oxford University Press, Oxford, UK, 3 edition, 1998.

17. J. Lin. Divergence measures based on the shannon entropy. *IEEE Trans. Inf. Theory*, 37(1):145–151, 1991.

18. R. A. Fisher. On the mathematical foundations of theoretical statistics. *Philos. Trans. Roy. Soc. A*, 222(594–604):309–368, 1922.

19. R. J. Rossi. *Mathematical statistics: An introduction to likelihood based inference.* John Wiley & Sons, Inc., New York, NY, 2018.

20. M. Rosenblatt. Remarks on some nonparametric estimates of a density function. *Ann. Math. Stat.*, 27(3):832–837, 1956.

21. E. Parzen. On estimation of a probability density function and mode. *Ann. Math. Stat.*, 33(3):1065–1076, 1962.

22. N. Privault. *Understanding Markov chains - Examples and Applications.* Springer undergraduate mathematics series. Springer, Singapore, 2 edition, 2018.

23. J. R. Norris. *Markov chains.* Cambridge series in statistical and probabilistic mathematics. Cambridge University Press, Cambridge, UK, 1997.

24. Carlo R. da Cunha. *Introduction to Econophysics: contemporary approaches with Python simulations.* CRC Press, Boca Raton, FL, 1st edition, 2021.

25. W. Zucchini, I. L. MacDonald, and R. Langrock. *Hidden Markov models for time series - An introduction using R.* Monographs on statistics and applied probability 150. CRC Press, Boca Raton, FL, 2nd edition, 2021.

26. L. E. Baum and T. Petrie. Statistical inference for probabilistic functions of finite state markov chains. *Ann. Math. Stat.*, 37(6):1554–1563, 1966.

27. L. R. Rabiner. A tutorial on hidden markov models and selected applications in speech recognition. *Proc. IEEE*, 77(1):257–286, 1989.

28. A. J. Viterbi. Error bounds for convolutional codes and an asymptotically optimum decoding algorithm. *IEEE Trans. Inf. Theory*, 13(2):260–269, 1967.

29. G. D. Forney, Jr. The viterbi algorithm. *Proc. IEEE*, 61(3):268–278, 1973.

30. J. L. W. V. Jensen. Sur les fonctions convexes et les inégalités entre les valeurs moyennes. *Acta Math.*, 30(1):175–193.

31. K. Pearson. On lines and planes of closest fit to systems of points in space. *Philos. Mag.*, 2(11):559–572, 1901.

32. H. Hotelling. Analysis of a complex of statistical variables into principal components. *J. Educ. Psychol.*, 24:417–441, 1933.

33. D. G. Luenberger and Y. Ye. *Linear and nonlinear programming.* International series in operations research & management science, 116. Springer, New York, NY, 3rd edition, 2008.

34. R. von Mises. *Mathematical Theory of Probability and Statistics.* Academic Press Inc., New York, NY, 1964.

35. C. Lanczos. An iteration method for the solution of the eigenvalue problem of linear differential and integral operators. *J. Res. Natl. Bur. Stand.*, 45(4):255–282, 1950.

36. B. Schölkopf and A. Smola. Nonlinear component analysis as a kernel eigenvalue problem. *Neural Comp.*, 10(5):1299–1319, 1998.

37. J. Mercer. Functions of positive and negative type, and their connection the theory of integral equations. *Philos. Trans. Roy. Soc. A*, 209(441-458):415–446, 1909.

38. J. V. Stone. Independent component analysis: a tutorial introduction. 2004.

39. A. Hyvärinen, J. Karhunen, and E. Oja. Independent component analysis. 2001.

40. A. J. Bell and T. J. Sejnowski. The "independent components" of natural scences are edge filters. *Vision Res.*, 37(23):3327–3338, 1997.

41. A. Hyvärinen. Fast and robust fixed-point algorithms for independent component analysis. *IEEE Trans. Neural Net.*, 10(3):626–634, 1999.

42. E. W. Forgy. Cluster analysis of multivariate data: effiency versus interpretability of classifications. *Biometrics*, 21(3):768–769, 1965.

43. L. Kaufman and P. J. Rousseeuw. Partitioning around the medoids (program pam). In *Finding groups in data: An introduction to cluster analysis*, Wiley series in probability and statistics, page 66–125. John Wiley & Sons, Inc., Hoboken, NJ, 1990.

44. P. J. Rousseeuw. Silhouettes: a graphical aid to the interpretation and validation of cluster analsysis. *Appl. Math.*, 20:53–65, 1987.

45. E. Simoudis, J. Han, and U. Fayyad, editors. *A density-based algorithm for discovering clusters in large spatial databases with noise*, 1996.

46. L. Kaufman and P. J. Rousseeuw. *Finding groups in data: an introduction to cluster analysis*. Wiley-Interscience, New York, NY, 1990.

47. C. D. Manning, P. Raghavan, and H. Schütze. *Introduction to information retrieval*. Cambridge University Press, Cambridge, UK, 2008.

48. M. Sokal. A statistical method for evaluating systematic relationships. *Univ. Kans. Sci. Bull.*, 38:1409–1438, 1958.

49. U. von Luxburg. A tutorial on spectral clustering. *Stat. Comput.*, 17:395–416, 2007.

50. M. Fiedler. Algebraic connectivity of graphs. *Czechoslov. Math. J.*, 23(98):298–305, 1973.

51. M. Fiedler. Laplacian of graphs and algebraic connectivity. *Banach Centr. Publ.*, 25(1):57–70, 1987.

52. T. Kohonen. *Self-Organizing Maps*. Springer Series in Information Sciences. Springer, Berlin, 3 edition, 1995.

53. T. Kohonen. Self-organized formation of topologically correct feature maps. *Bio. Cybernetics*, 43(1):59–69, 1982.

54. T. Martinetz and K. Schulten. A "neural-gas" network learns topologies. In T. Kohonen, K. Mäkisara, O. Simula, and J. Kangas, editors, *Artificial Neural Networks*. Elsevier (North-Holland), Amsterdam, 1st edition, 1991.

55. B. Fritzke. A growing neural gas network learns topologies. In G. Tesauro, D. S. Touretzky, and T. K. Leen, editors, *Advances in Neural Information Processing Systems 7*. MIT Press, Cambridge, ME, 1995.

56. J. de Loera, J. Rambau, and F. Santos. *Triangulations, Structures for Algorithms and Applications*. Algorithms and Computation in Mathematics - Volume 25. Springer-Verlag, Heidelberg, 1st edition, 2010.

57. S. P. Lloyd. Least square quantization in pcm. *IEEE Trans. Inf. Theory*, 28(2):129–137, 1982.

58. F. J. Anscombe. Graphs in statistical analysis. *Am. Stat.*, 27(1):17–21, 1973.

59. E. H. Simpson. The interpretation of interaction in contigency tables. *J. Royal Stat. Soc. B*, 13:238–241, 1951.

60. Ron Kohavi and David Wolpert. Bias plus variance decomposition for zero-one loss functions. In *Proceedings of the Thirteenth International Conference on International Conference on Machine Learning*, ICML'96, page 275–283, San Francisco, CA, USA, 1996. Morgan Kaufmann Publishers Inc.

61. N. R. Draper and H. Smith. *Applied regression analysis*. Wiley series in probability and statistics. Wiley-Interscience, New York, NY, 3 edition, 1998.

62. E. H. Moore. On the reciprocal of the general algebraic matrix. *Bull. Am. Math. Soc.*, 26(9):394–395, 1920.

63. R. Penrose. A generalized inverse for matrices. *Math. Proc. Camb. Philos. Soc.*, 51(3):406–413, 1955.

64. A. Gelman, J. B. Carlin, H. S. Stern, D. B. Dunson, A. Vehtari, and D. B. Rubin. *Bayesian data analysis*. Texts in statistical science. CRC Press, Boca Raton, FL, 2013.

65. A. N. Tikhonov, A. Goncharsky, V. V. Stepanov, and A. G. Yagola. *Numerical methods for the solution of ill-posed problems*. Mathematics and its applications, 328. Springer, New York, NY, 1995.

66. F. Santosa and W. W. Symes. Linear inversion of band-limited reflection seismograms. *SIAM J. Sci. Comput.*, 7(4):1307–1330, 1986.

67. R. Tibshirani. Regression shrinkage and selection via the lasso. *J. R. Stat. Soc. Series B Stat. Methodol.*, 58(1):267–288, 1996.

68. B. Efron, T. Hastie, I. Johnstone, and R. Tibshirani. Least angle regression. *Ann. Stat.*, 32(2):407–499, 2004.

69. K. Levenberg. A method for the solution of certain non-linear problems in least squares. *Q. Appl. Math.*, 2(2):164–168, 1944.

70. D. Marquardt. An algorithm for least-squares estimation of nonlinear parameters. *SIAM J. Appl. Math.*, 11(2):431–441, 1963.

71. C. E. Ramussen and C. K. I. Williams. Gaussian processes for machine learning. 2006.

72. D. J. C. MacKay. *Information theory, inference, and learning algorithms*. Cambridge University Press, Cambridge, UK, 2005.

73. Carl Edward Rasmussen. *Gaussian Processes in Machine Learning*, page 63–71. Springer Berlin Heidelberg, Berlin, Heidelberg, 2004.

74. L. Wasserman. *All of statistics - A concise course in statistical inference*. Springer texts in statistics. Springer, New York, NY, 1959.

75. P. F. Verhulst. Notice sur la loi que la population poursuit dans son accroisement. *Corresp. Math. Phys.*, 10:113–121, 1838.

76. S. H. Walker and D. B. Duncan. Estimation of the probability of an event as a function of several independent variables. *Biometrika*, 54(1/2):167–179, 1967.

77. J. R. Rice and K. H. Usow. The lawson algorithm and extensions. *Math. Comp.*, 22:118–127, 1968.

78. D. W. Hosmer, Jr., S. Lemeshow, and R. X. Sturdivant. *Applied logistic regression*. John Wiley & Sons, Inc., Hoboken, NJ, 3rd edition.

79. L. Breiman, J. H. Friedman, R. A. Olshen, and C. J. Stone. Classification and regression trees. *Biometrics*, 40(3):874, 1984.

80. J. R. Quinlan. Induction of decision trees. *Mach. Learn.*, 1:81–106, 1986.

81. L. Breiman. Bagging predictors. *Mach. Learn.*, 24(2):123–140, 1996.

82. T. K. Ho. The random subspace method for constructing decision forsts. *IEEE Trans. Pattern. Anal. Mach. Intell.*, 20(8):832–844, 1998.

83. Y. Amit and D. Geman. Shape quantization and recognition with randomized trees. *Neural Comp.*, 9(7):1545–1588, 1997.

84. Yoav Freund and Robert E. Schapire. A desicion-theoretic generalization of on-line learning and an application to boosting. In Paul Vitányi, editor, *Computational Learning Theory*, page 23–37, Berlin, Heidelberg, 1995. Springer Berlin Heidelberg.

85. K. G. Binmore. *Mathematical Analysis*. Cambridge University Press, Cambridge, UK, 2 edition, 1982.

86. C. Cortes and V. Vapnik. Support-vector networks. *Mach. Learn.*, 20:273–297, 1995.

87. A. Ruszczyński. *Nonlinear optimization*. Princeton University Press, Princeton, NJ, 2006.

88. J. Nocedal and S. J. Wright. *Numerical Optimization*. Springer Series in Operations Research. Springer, New York, NY, 2nd edition, 2006.

89. J. Platt. Sequential minimal optimization: A fast algorithm for training support vector machines. *Microsoft Research Technical Report*, 1998.

90. David J. Hand and Keming Yu. Idiot's Bayes: Not so stupid after all? *International Statistical Review / Revue Internationale de Statistique*, 69(3):385–398, 2001.

91. A. Kolmogorov. On the representation of continuous functions of several variables by superpositions of continuous functions of a smaller number of variables. *Proc. USSR Acad. Sci.*, 108:179–182, 1956.

92. V. Arnold. On functions of three variables. *Proc. USSR Acad. Sci.*, 28:51–54, 1963.

93. G. Cybenko. Approximations by superpositions of sigmoidal functions. *Math. Cont. Sig. Sys.*, 2(4):303–314, 1989.

94. K. Hornik. Approximation capabilities of multilayer feedforward networks. *Neural Networks*, 4(2):251–257, 1991.

95. W. McCulloch and W. Pitts. A logical calculus of ideas immanent in nervous activity. *Bull. Math. Bio.*, 5(4):115–133, 1943.

96. F. Rosenblatt. The perceptron - a perceiving and recognizing automaton. Technical report, 1957.

97. F. Rosenblatt. *Principles of neurodynamics - perceptrons and the theory of brain mechanisms.* Spartan Books, Washington D.C., 1962.

98. S. Linnainmaa. Taylor expansion of the accumulated rounding error. *BIT Numer. Math.*, 16(2):146–160, 1976.

99. V. E. Ismailov. *Ridge functions and applications in neural networks*, volume 263 of *Mathematical surveys and monographs*. American Mathematical Society, Providence, RI, 2021.

100. R. Pascanu, T. Mikolov, and Y. Bengio. On the difficulty of training recurrent neural networks. In S. Dasgupta and D. McAllester, editors, *Proceedings of the 30th International Conference on Machine Learning*, volume 28 of *Proceedings of Machine Learning Research*, page 1310–1318, Atlanta, Georgia, USA, 17–19 Jun 2013. PMLR.

101. H. Robbins and S. Monro. A stochastic approximation method. *Ann. Math. Statist.*, 22(3):400–407, 1951.

102. H. Robbins and D. Siegmund. A convergence theorem for non negative almost supermartingales and some applications**research supported by nih grant 5-r01-gm-16895-03 and onr grant n00014-67-a-0108-0018. In J. S. Rustagi, editor, *Optimizing Methods in Statistics*, page 233–257. Academic Press, 1971.

103. C. C. Aggarwal. *Linear Algebra and Optimization for Machine Learning.* Springer, Cham, Switzerland, 2020.

104. J. C. Spall. Introduction to stochastic search and optimization - estimation, simulation, and control. 2003.

105. I. Loshchilov and F. Hutter. SGDR: Stochastic gradient descent with warm restarts. In *International Conference on Learning Representations*, 2017.

106. D. E. Rumelhart, G. E. Hinton, and R. J. Williams. Learning representations by backpropagating errors. *Nature*, 323(6088):533–536, Oct 1986.

107. J. Duchi, E. Hazan, and Y. Singer. Adaptive subgradient methods for online learning and stochastic optimization. *J. Mach. Learn. Res.*, 12:2121–2159, 2011.

108. D. P. Kingma and J. Ba. Adam: A method for stochastic optimization. In Yoshua Bengio and Yann LeCun, editors, *3rd International Conference on Learning Representations, ICLR 2015, San Diego, CA, USA, May 7-9, 2015, Conference Track Proceedings*, 2015.

109. M. Innes, E. Saba, K. Fischer, D. Gandhi, M. C. Rudilosso, N. M. Joy, T. Karmali, A. Pal, and V. Shah. Fashionable modelling with flux. *CoRR*, abs/1811.01457, 2018.

110. M. Innes. Flux: Elegant machine learning with julia. *Journal of Open Source Software*, 2018.

111. X. Glorot and Y. Bengio. Understanding the difficulty of training deep feedforward neural networks. In Y. W. Teh and M. Titterington, editors, *Proceedings of the Thirteenth International Conference on Artificial Intelligence and Statistics*, volume 9 of *Proceedings of Machine Learning Research*, page 249–256, Chia Laguna Resort, Sardinia, Italy, 13–15 May 2010. PMLR.

112. K. He, X. Zhang, S. Ren, and J. Sun. Delving deep into rectifiers: Surpassing human-level performance on imagenet classification. In *2015 IEEE International Conference on Computer Vision (ICCV)*, page 1026–1034, 2015.

113. M. A. Kramer. Nonlinear principal component analysis using autoassociative neural networks. *AIChE Journal*, 37(2):233–243, 1991.

114. J. Schmidhuber. Deep learning in neural networks: an overview. *Neural Networks*, 61:85–117, 2015.

115. P. Vincent and H. Larochelle. Stacked denoising autoencoders: Learning useful representations in a deep network with a local denoising criterion. *J. Mach. Learn. Res.*, 11:3371–3408, 2010.

116. L. P. Cinelli, M. A. Marins, E. Antúnio B. da Silva, and S. L. Netto. *Variational Autoencoder*, page 111–149. Springer International Publishing, Cham, 2021.

117. D. M. Blei, A. Kucukelbir, and J. D. McAuliffe. Variational inference: A review for statisticians. *J. Am. Stat. Assoc.*, 112(518):859–877, 2017.

118. S. R. Bowman, L. Vilnis, O. Vinyals, A. Dai, R. Jozefowicz, and S. Bengio. Generating sentences from a continuous space. In *Proceedings of the 20th SIGNLL Conference on Computational Natural Language Learning*, page 10–21, Berlin, Germany, August 2016. Association for Computational Linguistics.

119. D. J. Rezende and S. Mohamed. Variational inference with normalizing flows. In *Proceedings of the 32nd International Conference on International Conference on Machine Learning - Volume 37*, ICML'15, page 1530–1538. JMLR.org, 2015.

120. J. Ho, A. Jain, and P. Abbeel. Denoising diffusion probabilistic models. In H. Larochelle, M. Ranzato, R. Hadsell, M.F. Balcan, and H. Lin, editors, *Advances in Neural Information Processing Systems*, volume 33, page 6840–6851. Curran Associates, Inc., 2020.

121. S. Gu, D. Chen, J. Bao, F. Wen, B. Zhang, D. Chen, L. Yuan, and B. Guo. Vector quantized diffusion model for text-to-image synthesis. In *2022 IEEE/CVF Conference on Computer Vision and Pattern Recognition (CVPR)*, page 10686–10696, 2022.

122. J. Sohl-Dickstein, E. A. Weiss, N. Maheswaranathan, and S. Ganguli. Deep unsupervised learning using nonequilibrium thermodynamics. In *Proceedings of the 32nd International Conference on International Conference on Machine Learning - Volume 37*, ICML'15, page 2256–2265. JMLR.org, 2015.

123. Y. Song, J. Sohl-Dickstein, D. P. Kingma, A. Kumar, S. Ermon, and B. Poole. Score-based generative modeling through stochastic differential equations. In *International Conference on Learning Representations*, 2021.

124. A. Hyvärinen. Estimation of non-normalized statistical models by score matching. *J. Mach. Learn. Res.*, 6:695–709, 2005.

125. M. Welling and Y. W. Teh. Bayesian learning via stochastic gradient langevin dynamics. In *Proceedings of the 28th International Conference on International Conference on Machine Learning*, ICML'11, page 681–688, Madison, WI, USA, 2011. Omnipress.

126. I. Goodfellow, J. Pouget-Abadie, M. Mirza, B. Xu, D. Warde-Farley, S. Ozair, A. Courville, and Y. Bengio. Generative adversarial nets. In Z. Ghahramani, M. Welling, C. Cortes, N. Lawrence, and K.Q. Weinberger, editors, *Advances in Neural Information Processing Systems*, volume 27. Curran Associates, Inc., 2014.

127. S. Boyd and L. Vandenberghe. *Convex Optimization*. Cambridge University Press, Cambridge, UK, 2004.

128. F. Brandt, M. Brill, and W. Suksompong. An ordinal minimax theorem. *Games Econ. Behav.*, 95:107–112, 2016.

129. J. von Neumann. Zur theorie der gesellschaftsspiele. *Math. Ann.*, 100:295–320, 1928.

130. D. H. Hubel and T. N. Wiesel. Receptive fields and functional architecture of monkey striate cortex. *J. Physiology*, 195(1):215–243, 1968.

131. K. Fukushima. Neocognitron: A self-organizing neural network model for a mechanism of pattern recognition unaffected by shift in position. *Bio. Cybernetics*, 36(4):193–202, 1980.

132. T. Homma, L. E. Atlas, and R. J. Marks II. An artificial neural network for spatio-temporal bipolar patterns: Application to phoneme classification. *Adv. Nerual Inf. Proc.*, 1:31–40, 1988.

133. A. V. Oppenheim, A. S. Willsky, and S. H. Nawab. *Signals & Systems*. Pearson, London, UK, 2 edition, 1996.

134. R. Venkatesan and B. Li. *Convolutional neural networks in visual computing: A concise guide*. Data-enabled engineering. CRC Press, Boca Raton, FL, 2018.

135. K. Fukushima. Visual feature extraction by a multilayered network of analog threshold elements. *IEEE Trans. Syst. Sci. Cybern.*, 5(4):322–333.

136. J. J. Weng, N. Ahuja, and T. S. Huang. Learning recognition and segmentation of 3-d objects from 2-d images. In *1993 (4th) International Conference on Computer Vision*, page 121–128, 1993.

137. H. T. Segelmann and E. D. Sontag. On the computational power of neural nets. *J. Comput. Syst. Sci.*, 50(1):132–150, 1995.

138. J. L. Elman. Finding structure in time. *Cogn. Sci.*, 14(2):179–211, 1990.

139. M. I. Jordan. Chapter 25 - serial order: A parallel distributed processing approach. In J. W. Donahoe and V. P. Dorsel, editors, *Neural-Network Models of Cognition*, volume 121 of *Advances in Psychology*, page 471–495. North-Holland, 1997.

140. P. J. Werbos. Generalization of backpropagation with application to a recurrent gas market model. *Neural Networks*, 1(4):339–356, 1988.

141. S. Hochreiter and J. Schmidhuber. Long short-term memory. *Neural Comp.*, 9(8):1735–1780, 1997.

142. I. Schlag, K. Irie, and J. Schmidhuber. Linear transformers are secretly fast weight programmers. In M. Meila and T. Zhang, editors, *Proceedings of the 38th International Conference on Machine Learning*, volume 139 of *Proceedings of Machine Learning Research*, page 9355–9366. PMLR, 18–24 Jul 2021.

143. T. Wolf, L. Debut, V. Sanh, J. Chaumond, C. Delangue, A. Moi, P. Cistac, T. Rault, R. Louf, M. Funtowicz, J. Davison, S. Shleifer, P. von Platen, C. Ma, Y. Jernite, J. Plu, C. Xu, T. Le Scao, S. Gugger, M. Drame, Q. Lhoest, and A. Rush. Transformers: State-of-the-art natural language processing. In *Proceedings of the 2020 Conference on Empirical Methods in Natural Language Processing: System Demonstrations*, page 38–45, Online, October 2020. Association for Computational Linguistics.

144. J. Schmidhuber. Learning to control fast-weight memories: an alternative to recurrent nets. *Neural Comp.*, 4(1):131–139, 1992.

145. A. Vaswani, N. Shazeer, N. Parmar, J. Uszkoreit, L. Jones, A. N. Gomez, Ł. Kaiser, and I. Polosukhin. Attention is all you need. In *Proceedings of the 31st International Conference on Neural Information Processing Systems*, NIPS'17, page 6000–6010, Red Hook, NY, USA, 2017. Curran Associates Inc.

146. K. He, X. Zhang, S. Ren, and J. Sun. Deep residual learning for image recognition. In *2016 IEEE Conference on Computer Vision and Pattern Recognition (CVPR)*, page 770–778, 2016.

147. J. J. Hopfield. Neural networks and physical systems with emergent collective computational abilities. *PNAS*, 79(8):2554–2558, 1982.

148. A. L. Hodgkin and A. F. Huxley. A quantitative description of membrane current and its application to conduction and excitation in nerve. *The Journal of Physiology*, 117(4):500–544, 1952.

149. W. Gerstner and W. M. Kistler. *Spiking neuron models*. Cambridge University Press, Cambridge, UK, 2002.

150. S. Löwel and W. Singer. Selection of intrinsic horizontal connections in the visual cortex by correlated neuronal activity. *Science*, 255(5041):209–212, 1992.

151. D. O. Hebb. *The organization of behavior, A neuropsychological theory.* Lawrence Erlbaum Associates, Mahwah, NJ, 2009.

152. E. Ising. Beitrag zur theorie des ferromagnetismus. *Zeitschrift für Physik*, 31(1):253–258, Feb 1925.

153. P. G. Drazin. Nonlinear systems. 1992.

154. R. J. McEliece, E. C. Posner, E. R. Rodemich, and S. S. Venkatesh. The capacity of the hopfield associative memory. *IEEE Trans. Inf. Theory*, IT-33(4):461–482, 1987.

155. D. Krotov and J. J. Hopfield. Dense associative memory for pattern recognition. In D. Lee, M. Sugiyama, U. Luxburg, I. Guyon, and R. Garnett, editors, *Advances in Neural Information Processing Systems*, volume 29. Curran Associates, Inc., 2016.

156. M. Demircigil, J. Heusel, M. Löwe, S. Upgang, and F. Vermet. On a model of associative memory with huge storage capacity. *J. Stat. Phys.*, 168:288–299, 2017.

157. H. Ramsauer, B. Schäfl, J. Lehner, P. Seidl, M. Widrich, L. Gruber, M. Holzleitner, T. Adler, D. Kreil, M. K. Kopp, G. Klambauer, J. Brandstetter, and S. Hochreiter. Hopfield networks is all you need. In *International Conference on Learning Representations*, 2021.

158. D. Krotov and J. J. Hopfield. Large associative memory problem in neurobiology and machine learning. In *International Conference on Learning Representations*, 2021.

159. A. L. Yuille and A. Rangarajan. The concave-convex procedure. *Neural Computation*, 15(4):915–936, 2003.

160. D. H. Ackley, G. E. Hinton, and T. J. Sejnowski. A learning algorithm for boltzmann machines*. *Cognitive Science*, 9(1):147–169, 1985.

161. D. Sherrington and S. Kirkpatrick. Solvable model of a spin-glass. *Phys. Rev. Lett.*, 35:1792–1796, Dec 1975.

162. G. E. Hinton. Training products of experts by minimizing contrastive divergence. *Neural Comp.*, 14(8):1771–1800, 2002.

163. G. Casella and E. I. George. Explaining the gibbs sampler. *The American Statistician*, 46(3):167–174, 1992.

164. I. P. Pavlov. *The work of the digestive glands.* HardPress Publishing, Miami, FL, 2 edition, 2015.

165. L. P. Kaelbling, M. L. Littman, and A. W. Moore. Reinforcement learning: A survey. *J. Artif. Intell. Res.*, 4:237–285, 1996.

166. R. Bellman. A markovian decision process. *J. Math. Mech.*, 6(5):679–684, 1957.

167. R. Bellman. The theory of dynamic programming. *Bull. Am. Math. Soc.*, 60(6):503–516, 1954.

168. R. Bellman. On the theory of dynamic programming. *PNAS*, 38(8):716–719, 1952.

169. R. S. Sutton and A. G. Barto. *Reinforcement learning*. The MIT Press, Cambridge, MA, 2 edition, 2020.

170. A. N. Burnetas and M. N. Katehakis. Optimal adaptive policies for markov decision processes. *Math. Oper. Res.*, 22:222–255, 1997.

171. R. S. Sutton. Learning to predict by the method of temporal differences. *Mach. Learn.*, 3:9–44, 1988.

172. M. N. Katehakis and A. F. Veinott. The multi-armed bandit problem: decomposition and computation. *Math. Oper. Res.*, 12(2):262–268, 1987.

173. T. H. Cormen, C. E. Leiserson, R. L. Rivest, and C. Stein. *Introduction to algorithms*. MIT Press, Cambridge, MA, 1 edition, 1990.

174. R. J. Williams. Simple statistical gradient-following algorithms for connectionist reinforcement learning. *Machine Learning*, 8(3):229–256, May 1992.

175. T. Blickle and L. Thiele. A comparison of selection schemes used in evolutionary algorithms. *Evolutionary Computation*, 4(4):361–394, 12 1996.

176. M. J. Kochenderfer, T. A. Wheeler, and K. H. Wray. *Algorithms for decision making*. The MIT Press, Cambridge, MA, 2022.

177. V. Mnih, A. Puigdomenech Badia, M. Mirza, A. Graves, T. Lillicrap, T. Harley, D. Silver, and K. Kavukcuoglu. Asynchronous methods for deep reinforcement learning. In M. F. Balcan and K. Q. Weinberger, editors, *Proceedings of The 33rd International Conference on Machine Learning*, volume 48 of *Proceedings of Machine Learning Research*, page 1928–1937, New York, New York, USA, 20–22 Jun 2016. PMLR.

178. W. Sun and Y. X. Yuan. *Optimization theory and methods: Nonlinear programming*. Springer optinmization and its Applications. Springer, New York, NY.

179. D. C. Sorensen. Newton's method with a model trust region modification. *SIAM Journal on Numerical Analysis*, 19(2):409–426, 1982.

180. J. Martens. New insights and perspectives on the natural gradient method. *J. Mach. Learn. Res.*, 21:1–76, 2020.

181. S. I. Amari. Natural gradient works efficiently in learning. *Neural Comp.*, 10(2):251–276, 1998.

182. J. Schulman, S. Levine, P. Moritz, M. Jordan, and Pieter Abbeel. Trust region policy optimization. In *Proceedings of the 32nd International Conference on International Conference on Machine Learning - Volume 37*, ICML'15, page 1889–1897. JMLR.org, 2015.

183. K. A. De Jong. *Evolutionary Computation - A unified Approach*. The MIT Press, Cambridge, MA, 2006.

184. A. M. Turing. I.—computing machinery and intelligence. *Mind*, LIX(236):433–460, 10 1950.

185. C. R. da Cunha, N. Aoki, D. K. Ferry, and Y.-C. Lai. A method for finding the background potential of quantum devices from scanning gate microscopy data using machine learning. *Machine Learning: Science and Technology*, 3(2):025013, jun 2022.

186. C. R. da Cunha, N. Aoki, D. K. Ferry, A. Velasquez, and Y. Zhang. An investigation of the background potential in quantum constrictions using scanning gate microscopy and a swarming algorithm. *Physica A: Statistical Mechanics and its Applications*, 614:128550, 2023.

187. C. Darwin. *The origin of species*. East India Publishing Company, Ottawa, Canada, 2022.

188. A. Holzinger, D. Blanchard, M. Bloice, K. Holzinger, V. Palade, and R. Rabadan. Darwin, lamarck, or baldwin: Applying evolutionary algorithms to machine learning techniques. In *Proceedings of the 2014 IEEE/WIC/ACM International Joint Conferences on Web Intelligence (WI) and Intelligent Agent Technologies (IAT) - Volume 02*, WI-IAT '14, page 449–453, USA, 2014. IEEE Computer Society.

189. B. L. Miller and D. E. Goldberg. Genetic algorithms, tournament selection, and the effects of noise. *Complex Systems*, 9:193–212, 1995.

190. D. E. Goldberg, B. Korb, and K. Deb. Messy genetic algorithms: Motivation, analysis, and first results. *Complex Systems*, 3:493–530, 1989.

191. A. Lipowski and D. Lipowska. Roulette-wheel selection via stochastic acceptance. *Physica A: Statistical Mechanics and its Applications*, 391(6):2193–2196, 2012.

192. J. E. Baker. Reducing bias and inefficiency in the selection algorithm. In *Proceedings of the Second International Conference on Genetic Algorithms on Genetic Algorithms and Their Application*, page 14–21, USA, 1987. L. Erlbaum Associates Inc.

193. H.-G. Beyer and H.-P. Schwefel. Evolution strategies – a comprehensive introduction. *Natural Computing*, 1(1):3–52, Mar 2002.

194. J. Kennedy and R. Eberhart. Particle swarm optimization. In *Proceedings of ICNN'95 - International Conference on Neural Networks*, volume 4, page 1942–1948, 1995.

195. M. Dorigo. *Optimization, Learning and Natural Algorithms (in Italian)*. PhD thesis, Dipartimento di Elettronica, Politecnico di Milano, Milan, Italy, 1992.

196. M. Dorigo, V. Maniezzo, and A. Colorni. Ant system: Optimization by a colony of cooperating agents. *IEEE Trans. Syst. Man Cybern. B*, 26(1):29–41, 1996.

197. M. Dorigo and L. M. Gambardella. Ant colony system: A cooperative learning approach to the traveling salesman problem. *IEEE Trans. Evol. Comp.*, 1(1):53–66, 1997.

198. T. Stützle and H. H. Hoos. Max-min ant system. *Future Gener. Comput. Syst.*, 16(8):889–914, 2000.

199. J. A. Nelder and R. Mead. A simplex method for function minimization. *Computer Journal*, 7(4):308–313, 1965.

200. N. Metropolis, A. W. Rosenbluth, M. N. Rosenbluth, A. H. Teller, and E. Teller. Equation of state calculations by fast computing machines. *J. Chem. Phys.*, 21(6):1087–1092, 12 2004.

201. W. K. Hastings. Monte carlo sampling methods using markov chains and their applications. *Biometrika*, 57(1):97–109, 04 1970.

202. S. Kirkpatrick, C. D. Gelatt, and M. P. Vecchi. Optimization by simulated annealing. *Science*, 220(4598):671–680, 1983.

203. G. Casella and R. L. Berger. *Statistics inference*. Brooks/Cole Cengage Learning, Belmont, CA, 2 edition, 2002.

204. A. G. Baydin, B. A. Pearlmutter, A. A. Radul, and J. M. Siskind. Automatic differentiation in machine learning: a survey. *J. Mach. Learn. Res.*, 18(153):1–43, 2018.

205. M. R. Hestenes and E. Stiefel. Methods of conjugate gradients for solving linear systems. *J. Res. Natl. Bur. Stand.*, 49(6):409–436, 1952.

206. T. Kloek and H. K. van Dijk. Bayesian estimates of equation system parameters: An application of integration by monte carlo. *Econometrica*, 46(1):1–19, 1978.

Index

Printed in the United States
by Baker & Taylor Publisher Services